Robot Operating System Cookbook

Over 70 recipes to help you master advanced ROS concepts

Kumar Bipin

BIRMINGHAM - MUMBAI

Robot Operating System Cookbook

Commissioning Editor: Gebin George
Acquisition Editor: Divya Poojari
Content Development Editor: Amrita Noronha
Technical Editors: Ishita Vora, Sneha Hanchate
Copy Editors: Safis Editing, Vikrant Phadkay
Project Coordinator: Shweta H Birwatkar
Proofreader: Safis Editing
Indexer: Tejal Daruwale Soni
Graphics: Jisha Chirayil
Production Coordinator: Aparna Bhagat

First published: June 2018

Production reference: 1280618

Published by Packt Publishing Ltd.
Livery Place
35 Livery Street
Birmingham
B3 2PB, UK.

ISBN 978-1-78398-744-3

www.packtpub.com

`mapt.io`

Mapt is an online digital library that gives you full access to over 5,000 books and videos, as well as industry leading tools to help you plan your personal development and advance your career. For more information, please visit our website.

Why subscribe?

- Spend less time learning and more time coding with practical eBooks and Videos from over 4,000 industry professionals

- Improve your learning with Skill Plans built especially for you

- Get a free eBook or video every month

- Mapt is fully searchable

- Copy and paste, print, and bookmark content

PacktPub.com

Did you know that Packt offers eBook versions of every book published, with PDF and ePub files available? You can upgrade to the eBook version at `www.PacktPub.com` and as a print book customer, you are entitled to a discount on the eBook copy. Get in touch with us at `service@packtpub.com` for more details.

At `www.PacktPub.com`, you can also read a collection of free technical articles, sign up for a range of free newsletters, and receive exclusive discounts and offers on Packt books and eBooks.

Contributors

About the author

Kumar Bipin has 15+ years of research and development experience with such world-famous consumer electronics companies as STMicroelectronics and Motorola. He has participated in the research fellowship program at SERC, IISc Bengaluru, earned MS in Robotics and Computer Vision from IIIT, Hyderabad. have been fosters confident in delivering system-level solutions for consumer electronics products, which includes system software, perception planning and control for autonomous vehicle. Currently, he is leading the Autonomous Vehicle Product- Autonomai at Tata Elxsi.

Firstly, I would like to thank my awesome wife, Dr. Ujjwala, for standing beside me throughout my career and while I was writing this book. I would like to thank the Packt team for their support while publishing my first book. Special thanks to Amrita, my ever-patient editor Finally, I would also like to mention my current company, Tata Elxsi, who provided consistent support and time for a few projects in this book.

About the reviewer

Jonathan Cacace was born in Naples, Italy, on December 13, 1987. He received his PhD in automation engineering from University of Naples Federico II. He is involved in several research projects focused on industrial and service robotics, in which he has developed several ROS-based applications integrating robot perception and control. He is the author of the second edition of *Mastering ROS for Robotics Programming*, published by Packt.

Packt is searching for authors like you

If you're interested in becoming an author for Packt, please visit authors.packtpub.com and apply today. We have worked with thousands of developers and tech professionals, just like you, to help them share their insight with the global tech community. You can make a general application, apply for a specific hot topic that we are recruiting an author for, or submit your own idea.

Table of Contents

Preface

Robot Operating Systems (ROS) is an open source middleware framework for the development of complex robotic systems and applications. Despite the fact that the research community is very active in developing applications with ROS and keeps adding to its features, the amount of reference material and documentation is not adequate for the amount of development being done. It is thus the purpose of this book to provide comprehensive and up-to-date information about ROS, which is presently acknowledged as the major development framework for robotics applications.

A brief introduction to the basics and foundations of ROS is addressed in the first few chapters in order to help beginners to get started. Advanced concepts will be dealt with later. First and foremost, the primary concepts around modeling and simulation in Gazebo/RotorS are discussed, which includes mobile robotics, micro aerial vehicles, and robotic arms—the three leading branches of robotic applications. Consequently, autonomous navigation frameworks for both mobile robots and micro aerial vehicles are introduced, which also includes the integration of ORB SLAM and PTAM. The book also covers programming of motion planning and grasping for robot manipulators.

Finally, the book discusses ROS-Industrial (ROS-I), an open source project that extends the advanced capabilities of ROS software to manufacturing industries.

I believe that this book will be a great guide that enables ROS users and developers to learn more about ROS's capabilities and features.

Who this book is for

The aim of this book is to provide an audience of engineers, whether from academia or industry, with a collection of problems, solutions, and future research issues they will encounter in the development of robotic applications. Basic knowledge of GNU/Linux, C++, and Python programming with a GNU/Linux environment is strongly recommended in order for the reader to easily comprehend the contents of the book.

What this book covers

Chapter 1, *Getting Started with ROS*, covers the installation of ROS on various platforms, including desktop systems, virtual machines, Linux containers, and ARM-based embedded boards.

Chapter 2, *ROS Architecture and Concepts – I*, explains the core concepts of ROS and how to work with the ROS framework.

Chapter 3, *ROS Architecture and Concepts – II*, discusses the advanced concepts of ROS, such as parameter serve, actionlib, pluginlib, nodelets, and Transform Frame (TF).

Chapter 4, *ROS Visualization and Debugging Tools*, discusses the various debugging and visualization tools available in ROS, such as gdb, valgrind, rviz, rqt, and rosbag.

Chapter 5, *Accessing Sensors and Actuators through ROS*, discusses interfacing hardware components, such as sensors and actuators, with ROS. It also covers interfacing sensors using I/O boards such as Arduino and Raspberry Pi.

Chapter 6, *ROS Modeling and Simulation*, introduces modeling of physical robots and simulation of virtual environments using Gazebo. Modeling of mobile robots and robotic arms is discussed.

Chapter 7, *Mobile Robot in ROS*, discusses one of the most powerful features of ROS—the Navigation Stack—which enables a mobile robot to move autonomously.

Chapter 8, *The Robotic Arm in ROS*, explains how to create and configure a MoveIt! package for a manipulator robot and perform motion planning and grasping.

Chapter 9, *Micro Aerial Vehicle in ROS*, introduces the Micro Aerial Vehicle (MAV) simulation framework (RotorS) to perform research and development on MAVs, which also includes autonomous navigation framework along with ORB SLAM and PTAM.

Chapter 10, *ROS-Industrial (ROS-I)*, discusses the ROS-Industrial package, which comes with a solution to interface with and control industrial robot manipulators. We use its powerful tools, such as MoveIt!, Gazebo, and RViz. This chapter also discusses the future of ROS-Industrial: hardware support, capabilities, and applications.

To get the most out of this book

Readers can work with almost all of the examples in the book using only a standard computer running Ubuntu 16.04/18.04, without any special hardware requirements. However, additional hardware components will be required while working with external sensors, actuators, and I/O boards.

Alternatively, readers can work with Ubuntu 16.04/18.04 installed on a virtual machine, such as Virtualbox or VMware hosted on a Windows system, although more computational power is required.

The robotic applications discussed in this book require commercial hardware such as I/O boards (Arduino, Odroid, and Raspberry Pi), perspective sensors (Kinect and camera), and actuator (Servomotor and Joystick).

Most importantly, it is recommended that reader should learn by experimenting with the source code provided in the book so that they become familiar with the technical concepts.

Download the example code files

You can download the example code files for this book from your account at www.packtpub.com. If you purchased this book elsewhere, you can visit www.packtpub.com/support and register to have the files emailed directly to you.

You can download the code files by following these steps:

1. Log in or register at www.packtpub.com.
2. Select the **SUPPORT** tab.
3. Click on **Code Downloads & Errata**.
4. Enter the name of the book in the **Search** box and follow the onscreen instructions.

Once the file is downloaded, please make sure that you unzip or extract the folder using the latest version of:

- WinRAR/7-Zip for Windows
- Zipeg/iZip/UnRarX for Mac
- 7-Zip/PeaZip for Linux

The code bundle for the book is also hosted on GitHub at
`https://github.com/PacktPublishing/Robot-Operating-System-Cookbook`. In case there's an update to the code, it will be updated on the existing GitHub repository.

We also have other code bundles from our rich catalog of books and videos available at
`https://github.com/PacktPublishing/`. Check them out!

Download the color images

We also provide a PDF file that has color images of the screenshots/diagrams used in this book. You can download it here: `http://www.packtpub.com/sites/default/files/downloads/RobotOperatingSystemCookbook_ColorImages.pdf`.

Conventions used

There are a number of text conventions used throughout this book.

`CodeInText`: Indicates code words in text, database table names, folder names, filenames, file extensions, pathnames, dummy URLs, user input, and Twitter handles. Here is an example: "Mount the downloaded `WebStorm-10*.dmg` disk image file as another disk in your system."

A block of code is set as follows:

```
<launch>
 <group ns="/">
  <param name="rosversion" command="rosversion roslaunch" />
  <param name="rosdistro" command="rosversion -d" />
  <node pkg="rosout" type="rosout" name="rosout" respawn="true"/>
 </group>
</launch>
```

Any command-line input or output is written as follows:

```
$ rostopic list
$ cd
```

Bold: Indicates a new term, an important word, or words that you see onscreen. For example, words in menus or dialog boxes appear in the text like this. Here is an example: "On the main toolbar, select **File** | **Open Workspace**, and choose the directory representing the ROS workspace."

Warnings or important notes appear like this.

 Tips and tricks appear like this.

Get in touch

Feedback from our readers is always welcome.

General feedback: Email `feedback@packtpub.com` and mention the book title in the subject of your message. If you have questions about any aspect of this book, please email us at `questions@packtpub.com`.

Errata: Although we have taken every care to ensure the accuracy of our content, mistakes do happen. If you have found a mistake in this book, we would be grateful if you would report this to us. Please visit `www.packtpub.com/submit-errata`, selecting your book, clicking on the Errata Submission Form link, and entering the details.

Piracy: If you come across any illegal copies of our works in any form on the Internet, we would be grateful if you would provide us with the location address or website name. Please contact us at `copyright@packtpub.com` with a link to the material.

If you are interested in becoming an author: If there is a topic that you have expertise in and you are interested in either writing or contributing to a book, please visit `authors.packtpub.com`.

Reviews

Please leave a review. Once you have read and used this book, why not leave a review on the site that you purchased it from? Potential readers can then see and use your unbiased opinion to make purchase decisions, we at Packt can understand what you think about our products, and our authors can see your feedback on their book. Thank you!

For more information about Packt, please visit `packtpub.com`.

Getting Started with ROS

1

In this chapter, we will discuss the following recipes:

- Installing ROS on desktop systems
- Installing ROS on a virtual machine
- Using ROS from a Linux container
- Installing ROS on an ARM-based board

Introduction

ROS is an open source software framework for programming robots and developing complex robotics applications. In addition, ROS is sometimes called a meta-operating system or middleware software framework because it performs many of the functions of an operating system, but it requires a host operating system, such as Linux.

Moreover, like any operating system, it provides a hardware abstraction layer to build robotics applications without worrying about the underlying hardware. The core of ROS is a message-passing middleware framework—synchronous or asynchronous—where processes and threads can communicate with and transport data between each other even when they are running from different machines. In addition, ROS software is organized as packages, offering good modularity and reusability, which can be integrated with and used for any custom robotic application with minimal changes.

The **Robot Operating System** (**ROS**) is a framework that is widely accepted and used in the robotics community, which ranges from researchers to professional developers of commercial robots. ROS was originally started in 2007 by the **Stanford Artificial Intelligence Laboratory** (**SAIL**) under the name SwitchYard, in support of the Stanford AI Robot project.

Since 2008, Willow Garage has been continuing the development, and recently ROS has been actively maintained by the **Open Source Robotics Foundation** (**OSRF**):

- **ROS**: http://www.ros.org/
- **OSRF**: http://www.osrfoundation.org/

Let's start with the introductory chapter of this book, where we will learn how to install ROS on various platforms.

Installing ROS on desktop systems

We assume that you have a desktop system with Ubuntu 16.04 or Ubuntu 18.04 **long-term support** (**LTS**) installed, with an Intel Corei5 processor @3.2 GHz and 8GB of RAM, or similar. Furthermore, it is necessary to have a basic knowledge and understanding of Linux and command tools. If you need to refresh your memory or learn about these tools, you can find a lot of relevant resources on the internet, or you can find books on these topics instead.

In this section, we will discuss the ROS distribution and the corresponding supported operating system, which will guide help us in selecting the right combination, depending upon our requirements.

ROS distribution

An ROS distribution is a versioned set of ROS packages. These are very similar to Linux distributions. The purpose of the ROS distributions is to let developers work against a relatively stable code base until they are ready to set up forward. Each distribution maintains a stable set of core packages up to the **end of life** (**EOL**) of the distribution.

The latest recommended ROS distribution is Kinectic Kame, which will get support up to May 2021, however, the latest ROS distribution is Melodic Morenia, which was released on 23rd May 2018, and will be supported up to May 2023. One of the problems with this latest ROS distribution is that most of the packages will not be available on it because it will take time to migrate them from the previous distribution, so we do not recommend this.

The list of distributions to use can be found on the ROS website
(`http://wiki.ros.org/distributions`):

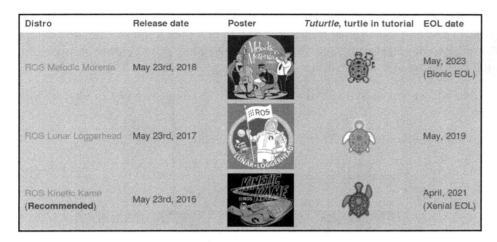

Distro	Release date	Poster	*Tuturtle*, turtle in tutorial	EOL date
ROS Melodic Morenia	May 23rd, 2018			May, 2023 (Bionic EOL)
ROS Lunar Loggerhead	May 23rd, 2017			May, 2019
ROS Kinetic Kame (Recommended)	May 23rd, 2016			April, 2021 (Xenial EOL)

List of distributions

Supported operating systems

ROS is fully compatible with the Ubuntu operating system by design. Moreover, the ROS
distributions are planned according to Ubuntu releases. Nevertheless, they are partially
supported by Ubuntu ARM, Gentoo, macOS X, Arch Linux, and OpenEmbedded. The
following table shows ROS distributions and the corresponding versions of the supported
operating systems:

ROS Distribution	Supported OS	Experimental OS
Melodic Morenia	Ubuntu 18.04 (LTS); Debian9	OS X (Homebrew), Gentoo, Ubuntu ARM, and OpenEmbedded/Yocto
Kinetic Kame (LTS)	Ubuntu 16.04 (LTS) and 15.10; Debian 8	OS X (Homebrew), Gentoo, Ubuntu ARM, and OpenEmbedded/Yocto
Jade Turtle	Ubuntu 15.04, 14.10, and 14.04; Debian 8	OS X (Homebrew), Gentoo, Arch Linux, Android NDK, Ubuntu ARM, and OpenEmbedded/Yocto
Indigo Igloo (LTS)	Ubuntu 14.04 (LTS) and 13.10; Debian 7	OS X (Homebrew), Gentoo, Arch Linux, Android NDK, Ubuntu ARM, and OpenEmbedded/Yocto

As we have discussed in the previous section, there are several ROS distributions available to download and install. Therefore, we recommend you select the LTS release, which is stable and getting maximum support.

However, if the latest features of ROS are required, you could go for the latest version, but you will not get the latest complete packages immediately after release because of the migration period from one distribution to another.

In this book, we are using two LTS distributions, ROS Kinetic and ROS Melodic, for all experiments.

The following screenshot shows a choice of ROS installations:

ROS installation

We can get the complete installation instructions for each distribution from the ROS website (`http://www.ros.org/`). Afterward, navigate to **Getting Started | Install**. It will show a screen listing, as shown in the preceding diagram, for the latest ROS distributions.

How to do it...

In the following section, we will go through the instructions for installing the latest ROS distribution.

Configuring Ubuntu repositories

To configure the Ubuntu repository, first search for **Software & Updates** in the Ubuntu Search Toolbar and enable the **restricted**, **universe**, and **multiverse** Ubuntu repositories, as shown in the following screenshot:

Software & Updates

Setting up the source.list file

The next step is to set up the desktop system to accept software from `packages.ros.org`. The details of the ROS repository server have to be added into `/etc/apt/source.list`:

```
$ sudo sh -c 'echo "deb http://packages.ros.org/ros/ubuntu $(lsb_release -
sc) main" > /etc/apt/sources.list.d/ros-latest.list'
```

Setting up keys

Whenever a new repository is added to the Ubuntu repository manager, we have to make it trusted to validate the origin of the packages by adding the keys. The following key should be added to Ubuntu before starting the installation, which will ensure the download comes from an authorized server:

```
$ sudo apt-key adv --keyserver hkp://ha.pool.sks-keyservers.net:80 --recv-
key 421C365BD9FF1F717815A3895523BAEEB01FA116
```

ROS Kinetic Installation

Now, we are prepared to install the ROS packages on Ubuntu. The first step is to update the list of packages by using the following command:

```
$ sudo apt-get update
```

There are many different libraries and tools in ROS, however, we will provide four default configurations to get you started:

- **Desktop-Full install** (recommended):

  ```
  $ sudo apt-get install ros-kinetic-desktop-full
  ```

- **Desktop install**:

  ```
  $ sudo apt-get install ros-kinetic-desktop
  ```

- ROS-Base:

  ```
  $ sudo apt-get install ros-kinetic-ros-base
  ```

- **Individual Package**:

  ```
  $ sudo apt-get install ros-kinetic-PACKAGE
  ```

ROS Melodic installation

Now, we are ready to install the ROS packages on Ubuntu 18.04 LTS. The first step is to update the list of packages by using the following command:

```
$ sudo apt-get update
```

There are many different libraries and tools in ROS, however, we will provide four default configurations to get you started:

- **Desktop-Full Install** (recommended):

  ```
  $ sudo apt-get install ros-melodic-desktop-full
  ```

- **Desktop Install**:

  ```
  $ sudo apt-get install ros-melodic-desktop
  ```

- **ROS-Base**:

  ```
  $ sudo apt-get install ros-melodic-ros-base
  ```

- **Individual Package**:

  ```
  $ sudo apt-get install ros-melodic-PACKAGE
  ```

Initializing rosdep

Before using ROS, you have to initialize rosdep, which enables you to easily install system dependencies for sources you want to compile, and also is required to run some core components in ROS:

```
$ sudo rosdep init
$ rosdep update
```

Setting up the environment

Good! We have completed the ROS installation. The ROS scripts and executables are mostly installed to /opt/ros/<ros_version>.

To get access to these scripts and executables, the ROS environment variables need to be added to the bash session. We have to source the following bash file for ROS Kinetic:

```
$ source /opt/ros/kinetic/setup.bash
```

ROS Melodic requires the following:

```
$ source /opt/ros/melodic/setup.bash
```

It's convenient if the ROS environment variables are automatically added to the bash session every time a new shell is launched.

For ROS Kinetic, use the following:

```
$ echo "source /opt/ros/kinetic/setup.bash" >> ~/.bashrc
$ source ~/.bashrc
```

For ROS Melodic, use the following:

```
echo "source /opt/ros/melodic/setup.bash" >> ~/.bashrc
source ~/.bashrc
```

If we have more than one ROS distribution installed, `~/.bashrc` must only source `setup.bash` for the version we are currently using:

```
$ source /opt/ros/<ros_version>/setup.bash
```

Getting rosinstall

Moreover, `rosinstall` is a frequently used command-line tool in ROS that is distributed separately. It enables us to easily download several source trees for the ROS packages with the single command.

This tool is based on Python, and can be installed using the following command:

```
$ sudo apt-get install python-rosinstall
```

Build Farm Status

The ROS packages are built by the ROS build farm. We can check the status of individual packages at `http://repositories.ros.org/status_page/`.

Congratulations! We are done with the ROS installation. We will execute the following command to check whether the installation is correct.

Open a new terminal to run `roscore`:

```
$ roscore
```

This is followed by a `turtlesim` node in another terminal:

```
$ rosrun turtlesim turtlesim_node
```

If everything is correct, we will get the following screen:

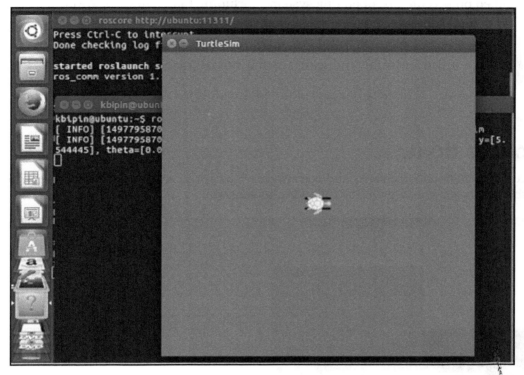

ROS execution demo

Installing ROS on a virtual machine

As we know, complete ROS support is only available on Ubuntu and Debian distributions. If we are Windows or macOS X users and we don't want to change the operating system of our computer to Ubuntu, we can use tools such as VMware or VirtualBox to help us to virtualize a new operating system on our computers.

VMware Workstation Pro is the industry standard for running multiple operating systems as virtual machines on a single PC. It is commercial software but also has free product trials and demos (https://www.vmware.com/in.html).

Alternatively, VirtualBox is a free and open source hypervisor for x86 computers that can be installed on several host operating systems, including Linux, macOS, Windows, Solaris, and OpenSolaris (`https://www.virtualbox.org/`).

We can get the detail information for VMware and VirtualBox from their official websites and search on the internet for tutorials instead.

How to do it...

Once the virtual machine starts and is configured properly, we should see another window on the host system, as seen in the following screenshot:

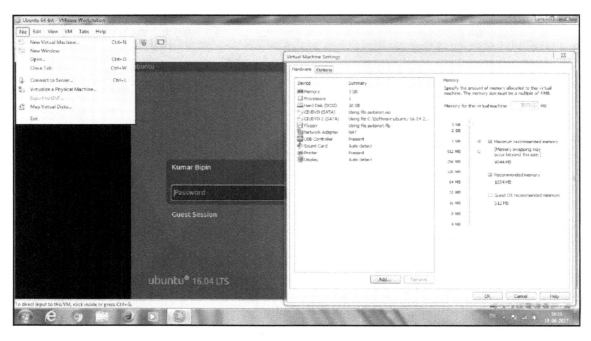

Virtual machine

There is no difference in an ROS installation on a virtual machine. Therefore, we can simply install ROS Kinetic following the same instructions described in the previous section. We can run most of the examples and stacks that we are going to work with. Unfortunately, the virtual machine may have problems when working and interfacing with external custom hardware through ROS. Moreover, performance degradation could also be observed with ROS running in a virtual machine. It is possible that the example source code discussed in Chapter 4, *ROS Visualization and Debugging Tools*, will not work.

The following screenshot shows ROS running on one of the virtual machines and the ROS installation:

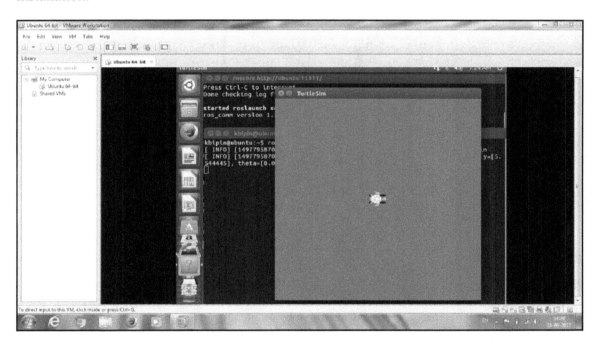

ROS on virtual machine

Using ROS from a Linux container

As the industry moves beyond the virtual machine consolidation paradigm, several types of containers have come into prominence. Two flavors currently enjoy the lion's share of deployments on the Linux operating system: Docker and LXC.

Getting ready

LXC (Linux Containers) is an OS-level virtualization technology that allows the creation and running of multiple isolated Linux **virtual environments** (**VE**) on a single control host. These isolation levels or containers can be used to either sandbox specific applications or emulate an entirely new host. LXC uses the Linux cgroups functionality, which was introduced in version 2.6.24, to allow the host CPU to better partition memory allocation into isolation levels called **namespaces** (https://linuxcontainers.org/).

Docker, previously called **dotCloud**, was started as a side project and only open sourced in 2013. It is really an extension of the LXC capabilities. This it achieved using a high-level API that provides a lightweight virtualization solution to run processes in isolation. Docker is developed in the Go language and utilizes LXC, cgroups, and the Linux kernel itself. Since it's based on LXC, a Docker container does not include a separate operating system; instead it relies on the operating system's own functionality, as provided by the underlying infrastructure. So, Docker acts as a portable container engine, packaging the application and all of its dependencies in a virtual container that can run on any Linux server (https://www.docker.com/).

 Note that a Linux VE is distinct from a **virtual machine** (**VM**).

How to do it...

Since Docker is an open platform that helps to distribute applications and complete systems and is based on LXC, we will discuss working with ROS Docker Images and how to distribute complex applications along with complete systems as a standalone image.

Installing Docker

Before installing Docker, it will require the updated packages:

```
$ sudo apt-get update
```

Use the following command to add the GPG key for the official Docker repository to the system:

```
$ sudo apt-key adv --keyserver hkp://p80.pool.sks-keyservers.net:80 --recv-keys 58118E89F3A912897C070ADBF76221572C52609D
```

Adding the Docker repository to APT sources

For Ubuntu 16.04, add the following:

```
$ echo "deb https://apt.dockerproject.org/repo ubuntu-xenial main" | sudo tee /etc/apt/sources.list.d/docker.list
```

For Ubuntu 18.04, add the following:

```
$ echo "deb https://apt.dockerproject.org/repo ubuntu-xenial main" | sudo tee /etc/apt/sources.list.d/docker.list
```

Updating the package database with the Docker packages from the newly added repository is done as follows:

```
$ sudo apt-get update
```

Make sure that we are about to install from the Docker repository instead of the default Ubuntu repository:

```
$ apt-cache policy docker-engine
```

Notice that docker-engine is not installed yet; to install docker-engine, use the following command:

```
$ sudo apt-get install -y docker-engine
```

Check whether Docker is started or not:

```
$ sudo systemctl status docker
```

To start the Docker service, use the following command:

```
$ sudo service docker start
$ docker
```

We can search for images available on Docker Hub by using the docker command with the search subcommand:

```
$ sudo docker search Ubuntu
```

To run the Docker container, use the following command:

```
$ sudo docker run -it hello-world
```

Getting and using ROS Docker images

Docker images are akin to packages such as virtual machines or complete systems that are already set up. There are servers that provide the images, and users only have to download them. The main server is Docker Hub, which is located at https://hub.docker.com. Here, it is possible to search for Docker images for different systems and configurations.

Well! All ROS Docker images are listed in the official ROS repository on the web at https://hub.docker.com/_/ros/. We will use ROS Kinetic images, which are already available here:

```
$ sudo docker pull ros
$ sudo docker pull kinetic-ros-core
$ sudo docker pull kinetic-ros-base
$ sudo docker pull kinetic-robot
$ sudo docker pull kinetic-perception
```

Similarly, we could pull ROS Melodic images, which are also already available there:

```
$ sudo docker pull ros
$ sudo docker pull melodic-ros-core
$ sudo docker pull melodic-ros-base
$ sudo docker pull melodic-robot
$ sudo docker pull melodic-perception
```

After the container is downloaded, we can run it interactively with the following command:

```
$ docker run -it ros
```

This will be like entering a session inside the Docker container. The preceding command will create a new container from the main image wherein we have a full Ubuntu system with ROS Kinetic already installed. We will be able to work as in a regular system, install additional packages, and run the ROS nodes.

We can list all the Docker containers available and the origin of their images:

```
$ sudo docker ps
```

Congratulations! We have completed the Docker installation. To set up the ROS environment inside the container in order to start using ROS, we have to run the following command (`ros_version: kinetic/melodic`):

```
$ source /opt/ros/<ros_version>/setup.bash
```

However, in principle, running docker should be enough, but we could even SSH into a running Docker container as a regular machine, using name or ID.

```
$ sudo docker attach 665b4a1e17b6 #by ID ...OR
$ sudo docker attach loving_heisenberg #by Name
```

If we use `attach`, we can use only one instance of the shell. So, if we want to open a new terminal with a new instance of a container's shell, we just need to run the following:

```
$ sudo docker exec -i -t 665b4a1e17b6 /bin/bash #by ID ...OR
$ sudo docker exec -i -t loving_heisenberg /bin/bash #by Name
```

Moreover, Docker containers can be stopped from other terminals using `docker stop`, and they can also be removed with `docker rm`.

See also

This book comes with a working Docker Image, which is basically an extension of ROS Kinetic with the code for the examples. The instructions to download and install it will be on the GitHub repository with the rest of the code.

Installing ROS on an ARM-based board

There is high demand from industries that are working on commercial robotic products for ROS that runs on an embedded platform, mostly on the ARM platform. Finally, OSRF, which maintains the open source ROS, has announced its formal support for an ARM target.

Getting ready

There are two streams to support ROS on an ARM target:

- Ubuntu ARM

- OpenEmbedded (meta-ros)

Ubuntu ARM is the most popular among researchers since it is easy to install and a lot of ARM boards are already supporting it; a few of them are shown in the following diagram. In addition, many ROS packages are already supported or could be ported with minimal changes:

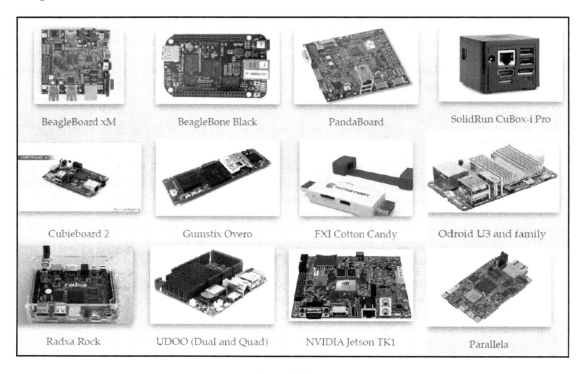

Supported platforms

However, OpenEmbedded is used by professional developers for commercial products in industries. The following table shows a comparison of both:

Ubuntu ARM	OpenEmbedded
Binary ROS packages	A cross-compilation tool chain for ROS packages based on catkin
Is compiled for a generic ARM architecture	Compiles all packages from their source
Installation with usual Ubuntu tools (dpkg, APT, and so on)	Supports many architectures: ARM, MIPS, PowerPC, and more
Easy and quick installation	Easy to adjust to new machines and architectures
No need to compile the basic ROS packages from source	Allows changes to the basic ROS packages
Common Ubuntu feel	Small Linux kernels and images
Additional compilation is onboard	Requires a powerful system setup to get the build machine and tool chain running

Despite several possibilities, we have chosen to install ROS on Ubuntu ARM because these distributions are more common and can be used on other ARM-based boards such as UDOO, ODROID-U3, ODROID-X2, or Gumstix Overo. It is recommended to use an image of Ubuntu ARM 16.04 Xenial armhf on the platform to work with ROS.

Before installing ROS on the specific ARM platform, we have to complete a few prerequisites. As this book is focused on ROS, we will list them without going into detail. However, there is a lot of information about Ubuntu ARM for specific ARM platforms available on websites, forums, and books that could be reviewed.

When we have Ubuntu ARM on our selected platform, the network interfaces must be installed to provide access to the network by configuring the network settings, such as the IP, DNS, and gateway.

An Ubuntu image for most of the ARM platform will be set up for MicroSD cards with 1-4 GB size. This is not sufficient to use a large part of the ROS Kinetic packages. In order to solve this problem, we can use SD cards with more space and expand the file system to occupy all the space available with re-partitioning:

GParted

We could use the GParted utility, an open source graphical tool that is used for creating, deleting, resizing, moving, checking, and copying disk partitions and their file systems (http://gparted.org/).

How to do it...

Good! We should be ready to install ROS. After this, the process of installation is pretty similar to the desktop installation discussed in the previous section. The major difference when installing ROS on the Ubuntu ARM platform is that it will not be possible to go for the full desktop installation. We install the selected package that is required for our application. Nevertheless, it will be nice to work with source building and installation for a package not present in the ROS repository:

- ROS Kinectic `<ros_version>` is compatible with Ubuntu 16.04 Xenial Xerus

- ROS Melodic `<ros_versions>` is compatible with Ubuntu 18.04 Bionic Beaver

Configuring repositories

The first step consists of configuring our Ubuntu repositories to allow "restricted," "universe," and "multiverse":

```
$ sudo vi /etc/apt/sources.list
```

We have something like this for Ubuntu 16.04:

```
deb http://ports.ubuntu.com/ubuntu-ports/ xenial main restricted universe
multiverse
#deb-src http://ports.ubuntu.com/ubuntu-ports/ xenial main restricted
universe multiverse deb http://ports.ubuntu.com/ubuntu-ports/ xenial-
updates main restricted universe multiverse
#deb-src http://ports.ubuntu.com/ubuntu-ports/ xenial-updates main
restricted universe multiverse
#Kernel source (repos.rcn-ee.com) :
https://github.com/RobertCNelson/linux-stable-rcn-ee
#
#git clone https://github.com/RobertCNelson/linux-stable-rcn-ee
#cd ./linux-stable-rcn-ee #git checkout `uname -r` -b tmp
# deb [arch=armhf] http://repos.rcn-ee.com/ubuntu/ xenial main
#deb-src [arch=armhf] http://repos.rcn-ee.com/ubuntu/ xenial main
```

Similarly, the following is for Ubuntu 18.04:

```
deb http://ports.ubuntu.com/ubuntu-ports/ bionic main restricted universe
multiverse
#deb-src http://ports.ubuntu.com/ubuntu-ports/ bionic main restricted
universe multiverse deb http://ports.ubuntu.com/ubuntu-ports/ bionic-
updates main restricted universe multiverse
#deb-src http://ports.ubuntu.com/ubuntu-ports/ bionic-updates main
```

```
restricted universe multiverse
#Kernel source (repos.rcn-ee.com) :
https://github.com/RobertCNelson/linux-stable-rcn-ee
#
#git clone https://github.com/RobertCNelson/linux-stable-rcn-ee
#cd ./linux-stable-rcn-ee
#git checkout `uname -r` -b tmp
# deb [arch=armhf] http://repos.rcn-ee.com/ubuntu/ bionic main
#deb-src [arch=armhf] http://repos.rcn-ee.com/ubuntu/ bionic main
```

Then, use the following to update the sources:

```
$ sudo apt-get update
```

Setting system locale

Some ROS tools, such as Boost, require that the system locale be set. This can be set with the following:

```
$ sudo update-locale LANG=C LANGUAGE=C LC_ALL=C LC_MESSAGES=POSIX
```

Setting up sources.list

Next, we will configure the source lists depending on the Ubuntu version installed in our ARM platform. Run the following command to install the Ubuntu armhf repositories:

```
$ sudo sh -c 'echo "deb http://packages.ros.org/ros/ubuntu $(lsb_release -cs) main" > /etc/apt/sources.list.d/ros-latest.list'
```

Setting up keys

As discussed previously, this step is needed to confirm that the origin of the code is correct and that no one has modified the code or programs without the knowledge of the owner:

```
$ sudo apt-key adv --keyserver hkp://ha.pool.sks-keyservers.net --recv-key 0xB01FA116
```

Installing the ROS packages

Before the installation of the ROS packages, make sure our Debian package index is up-to-date:

```
$ sudo apt-get update
```

There are many different libraries and tools in ROS—not all compile fully on ARM. So, it is not possible to make a full desktop installation. We should install ROS packages individually.

We could install `ros-base` (Bare Bones), which includes the ROS package, build, and communication libraries, but does not include GUI tools (press *ENTER* (Y) when prompted):

```
$ sudo apt-get install ros-<ros_version>-ros-base
```

However, we could try to install the desktop installation, which includes the ROS, rqt, RViz, and robot-generic libraries:

```
$ sudo apt-get install ros-<ros_version>-desktop
```

Adding individual packages

We can install a specific ROS package:

```
$ sudo apt-get install ros-<ros_version>-PACKAGE
```

Find available packages with the following command:

```
$ apt-cache search ros-<ros_version>
```

Initializing rosdep

The `rosdep` command-line tool must be installed and initialized before we can use ROS. This allows us to easily install libraries and solve system dependencies for the source we want to compile, and is required to run some core components in ROS:

```
$ sudo apt-get install python-rosdep $ sudo rosdep init $ rosdep update
```

Environment setup

Well done! We have completed the ROS installation for the ARM platform. The ROS scripts and executables are mostly installed in /opt/ros/<ros_version>.

To get access to these scripts and executables, the ROS environment variables need to be added to the bash session. We have to source the following bash file:

```
$ source /opt/ros/<ros_version>/setup.bash
```

It's convenient if the ROS environment variables are automatically added to the bash session every time a new shell is launched:

```
$ echo "source /opt/ros/<ros_version>/setup.bash" >> ~/.bashrc
$ source ~/.bashrc
```

If we have more than one ROS distribution installed, ~/.bashrc must only source the setup.bash for the version we are currently using:

```
$ source /opt/ros/<ros_version>/setup.bash
```

Getting rosinstall

rosinstall is a frequently used command-line tool in ROS that is distributed separately. It enables us to easily download several source trees for the ROS packages with a single command.

This tool is based on Python and can be installed using the following command:

```
$ sudo apt-get install python-rosinstall
```

As a basic example, we could run an ROS core on one terminal:

```
$ roscore
```

And from another terminal, we can publish a pose message:

```
$ rostopic pub /dummy geometry_msgs/Pose Position: x: 3.0 y: 1.0 z: 2.0
Orientation: x: 0.0 y: 0.0 z: 0.0 w: 1.0 -r 8
```

Moreover, we could set ROS_MASTER_URI on our desktop system (in the same network) to point to our ARM Platform (IP 192.168.X.X).

On your laptop, add the following:

```
$ export ROS_MASTER_URI=http://192.168.1.6:11311
```

And, we will see pose published from the ARM platform to our laptop.

On your laptop, add the following:

```
$ rostopic echo -n2 /dummy Position: x: 1.0 y: 2.0 z: 3.0 Orientation: x:
0.0 y: 0.0 z: 0.0 w: 1.0 ---
```

2
ROS Architecture and Concepts I

In this chapter, we will discuss the following recipes:

- Exploring the ROS filesystem
- Analyzing the ROS computation graph
- Associating with the ROS community
- Learning working with ROS

Introduction

In the previous chapter, we learned how to install ROS. Subsequently, in this chapter, we will look into the ROS architecture and concepts along with its components. Furthermore, we will learn how to create and use the ROS components—nodes, packages messages, services, and much more, with examples.

The ROS architecture and design has been divided into three sections or levels of concepts:

- **The ROS filesystem**: In this level, a group of concepts is used to explain how ROS is internally formed, the folder structure, and the minimum number of files that it needs to work.
- **The ROS computation graph**: In this section, we will see all the concepts and mechanisms which are required to set up the ROS computational network and environment, handle all the processes, and communicate with more than a single computer, and so on.

- **The ROS community**: This level is of great importance; as with most open source software projects, it comprises a set of tools and concepts to share knowledge, algorithms, and code among developers. Moreover, having a strong community not only helps newcomers to understand the intricacies of the technology, as well as solve the most common issues, it is also the main force that drives its growth.

Exploring the ROS filesystem

The main purpose of the ROS filesystem is to centralize the build process of a project, while at the same time providing enough flexibility to decentralize its dependencies. Nevertheless, the ROS filesystem is one of the important concepts to understand while developing projects in ROS, but, with time and patience, the developer will easily become familiar with it and realize its usefulness in managing complex projects and their dependencies:

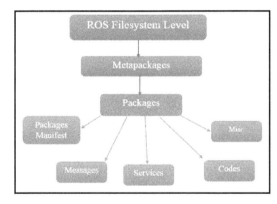

ROS filesystem

Getting ready

Similar to an operating system, an ROS filesystem is divided into folders, and these folders have files that describe their functionalities:

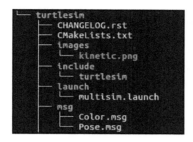

ROS package

- **ROS Packages**: The ROS software is organized in the form of packages, as a primary unit. A package might contain ROS nodes (processes), ROS-independent library, configuration files and so on, which could be logically define as a complete software module. For example, `chapter2_tutorials` is one of the simple packages.

- **The ROS Package Manifest**: The package manifest is an XML file called `package.xml`, which must be included in the package folder. This file contains information about the package, such as the package name, version, authors, maintainers, license, compilation flags, and dependencies on other packages. The system package dependencies are also declared in `package.xml`, which is essential to build any package from the source on any given machine.

- **ROS Messages**: In the ROS framework, nodes communicate with each other asynchronously by publishing messages to topics, which is defined as a simple data structure. The message file consists of typed fields and is placed inside the `msg` folder of a package that has the file extension `.msg`.

- **ROS Services**: ROS nodes can also exchange a request and response message as part of a ROS service call synchronously. These request and response messages are defined as `.srv` files inside the `srv` folder of a package.

- **The ROS Metapackage Manifests**: Although the metapackage manifests `package.xml`, are similar to the package manifest, they are specialized packages in the ROS catkin build system, which does not contain any code, files, or other items except the `package.xml` manifest. A metapackage defines the virtual packages used in the Debian packaging world and provides references to one or more related packages grouped together.

- **The ROS Metapackages**: It combines several packages in a group for a special purpose and functionality, such as a navigation task. However, in older versions of ROS such as Electric and Fuerte, it was called stacks, but later it was removed and metapackages came into existence. The metapackages could be seen as a simpler and easy representation of package stack. One of the examples of a meta package is the ROS navigation stack.

How to do it...

In this section, we will practice what we have learned until now. We will learn about ROS tools for navigating through the ROS filesystem and work with examples to practice setting up the ROS Workspace and the creation and building of a package and the metapackage. For creating the ROS workspace, we will perform the following steps:

1. First, we will create our own workspace. In this workspace, we will centralize all the example source code used throughout this cookbook.

2. Get the workspace that ROS is using, using the following command:

   ```
   $ echo $ROS_PACKAGE_PATH
   ```

   This will show an output similar to the following:

   ```
   /opt/ros/kinetic/share:/opt/ros/kinetic/stacks
   ```

3. Create and initialize the ROS workspace:

   ```
   $ mkdir -p ~/catkin_ws/src
   $ cd ~/catkin_ws/src
   $ catkin_init_workspace
   ```

   We have created an empty ROS workspace, which contains only a CMakeList.txt file.

4. To build the workspace, write the following commands:

   ```
   $ cd ~/catkin_ws
   $ catkin_make
   ```

   We can see that the new folders, build and devel, were created with the previous make command.

5. To finish the configuration, use the following command:

   ```
   $ source devel/setup.bash
   ```

   This step will reload the `setup.bash` file from the ROS workspace and set up an overlay over the default configuration.

6. To get the same environment for every shell on opening and closing, the following should be added in the `~/.bashrc` scripts:

   ```
   $ echo "source /opt/ros/kinetic/setup.bash" >> ~/.bashrc
   $ echo "source /home/usrname/catkin_ws/devel/setup.bash" >>
   ~/.bashrc
   ```

   ROS provides a number of command-line tools for navigating through the filesystem. We will look into some of most used ones.

   To get information about the packages and stacks in the ROS environment, such as their paths, dependencies, and so on, `rospack` and `rosstack` can be used. Similarly, to move through packages and stacks, as well as listing their contents, `roscd` and `rosls` can be used.

   For example, the path of the `turtlesim` package can be found with the following command:

   ```
   $ rospack find turtlesim
   ```

   This will result in the following output:

   ```
   /opt/ros/kinetic/share/turtlesim
   ```

   A similar result will be shown for installed metapackages with the following command:

   ```
   $ rosstack find ros_comm
   ```

7. To get the path for the `ros_comm` metapackage, use the following:

   ```
   /opt/ros/kinetic/share/ros_comm
   ```

   The following command lists the files inside the package or metapackage:

   ```
   $ rosls turtlesim
   ```

   The output of the preceding command will be as follows:

   ```
   cmake     images     srv     package.xml  msg
   ```

8. The current working directory can be changed with `roscd`:

```
$ roscd turtlesim
$ pwd
```

The new working directory will be as follows:

```
/opt/ros/kinetic/share/turtlesim
```

In ROS, package creation can be done manually, but to avoid the complex and tedious work involved, it is recommended to use the `catkin_create_pkg` command-line tool instead.

We will create a new package in our recently initialized workspace using the following commands:

```
$ cd ~/catkin_ws/src
$ catkin_create_pkg chapter2_tutorials std_msgs roscpp
```

This command includes the name of the package and the dependencies, in our case, `std_msgs` and `roscpp`.

The syntax of the command is as follows:

```
catkin_create_pkg [package_name] [dependency1] ... [dependencyN]
```

The `std_msgs` package contains common message types representing primitive data types and other basic message constructs, such as multiarray. The
`roscpp` package is a C++ implementation of ROS. It provides a client library that enables C++ programmers to quickly interface with ROS topics, services, and parameters.

As discussed earlier, we can use the `rospack`, `roscd`, and `rosls` commands to retrieve information about the ROS package.

- `rospack profile`: It informs us about the newly-added packages to the ROS system, which is useful after installing any new package.
- `rospack find chapter2_tutorials`: It helps us in finding the path of the package.
- `rospack depends chapter2_tutorials`: It retrieves the package dependencies.

- `rosls chapter2_tutorials`: It shows the package content.
- `roscd chapter2_tutorials`: This command changes the working directory.

Although the creation of a metapackage has similar steps to those for creating a package, it has a few limitations, and special cases, as discussed in the previous section so, should be handled with care. Once we have our package created and have written some code, it is necessary to build the package. The build process for a package compiles not only the code added by the user but also the code generated from the messages and services.

9. To build a package, we will use the `catkin_make` tool, as follows:

```
$ cd ~/catkin_ws/
$ catkin_make
```

The `catkin_make` command should be run in the workspace folder. If it were run in any other folder, eventually the command would fail. However, `catkin_make` is executed in the `catkin_ws` folder, and we will get a complete successful compilation. Moreover, a compilation of a single package using `catkin_make` could be done using the following command:

```
$ catkin_make --pkg <package name>
```

There's more...

In this section, we will provide more insight into ROS filesystem terminology, which includes the ROS workspace, package and metapackage, messages, and services. An experienced reader could jump to the next section, but it is recommended to run through this for a complete overview:

- **The ROS workspace**: Practically, the workspace is a folder that contains packages, which contain source files and configuration information. Moreover, the workspace provides the mechanism for centralizing all of the development.

A typical workspace is shown in the following screenshot:

```
 └── catkin_ws
     ├── build
     │   ├── catkin
     │   ├── catkin_generated
     │   ├── Makefile
     │   └── ...
     ├── devel
     │   ├── setup.zsh
     │   └── ...
     └── src
         ├── CMakeLists.txt -> /opt/ros/kinetic/share/catkin/cmake/toplevel.cmake
         └── ...
```

An explanation of the preceding workspace follows:

- `src`: The source space `src`, contains packages, projects, clone packages, and so on. One of the most important files in this space is `CMakeLists.txt`, which is invoked by `cmake` when packages are configured in the workspace. Moreover, this file is created with the `catkin_init_workspace` command during workspace initialization.
- `build`: In the build space, `build`, `cmake`, and `catkin` make tools keep the cache information, configuration, and other intermediate files for intended packages and projects.
- `devel`: The development space, `devel`, is used to keep the compiled packages, which can be tested without the installation step.

Another interesting feature of the ROS workspace is its overlays feature. When we are working with any package of ROS, for example, turtlesim, we could have a precompiled installed version, or we could download the source file and compile it to use as a development version.

ROS permits us to use the development version of any package instead of the installed version. This is a very useful feature when we are working on an upgrade of an installed package. We might not understand the overlay utility at this moment, but, in the following chapters, we will use this feature to create our own plugins.

Usually, when we talk about packages, we refer to a typical structure of files and folders in the ROS system. This structure looks as follows:

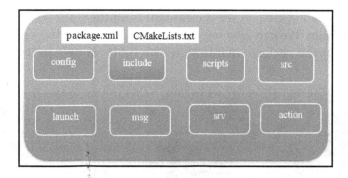

The structure of a typical ROS package contains the following:

- config: All configuration files that are used in this ROS package are kept in this folder. This folder is created by the user and it is a common practice to name the folder config to keep the configuration files in it.
- include: This folder consists of the headers of the libraries that need to be used inside the package.
- scripts: These are executable scripts that can be in Bash, Python, or any other scripting language.
- src: This is where the source files of our programs are present. We can create a folder for nodes and nodelets or organize as per convenient.
- launch: This folder keeps the launch files that are used to launch one or more ROS nodes.
- msg: This folder contains custom message type definitions.
- srv: This folder contains the service type definitions.
- action: This folder contains the action definition. We will look into more about actionlib in the upcoming sections.
- package.xml: This is the package manifest file of this package.
- CMakeLists.txt: This is the CMake build file of this package.

Metapackages are specialized packages in ROS that only contain one file, package.xml. It simply groups a set of multiple packages as a single logical package. In the package.xml file, the meta package contains an export tag. The ROS navigation stack is a good example of metapackages. We can locate the navigation metapackage using the following command:

```
$ rosstack find navigation $ roscd navigation $ gedit package.xml
```

In the following screenshot, we can see the content from the `package.xml` file in the navigation metapackage. We can also see the `<export>` tag and the `<run_depend>` tag. These are necessary in the package manifest, which is also shown in the following screenshot:

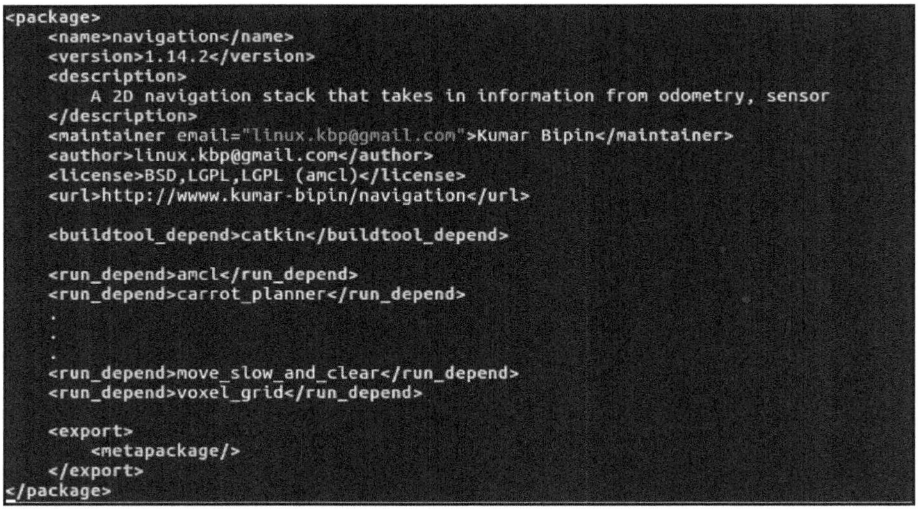

```
<package>
    <name>navigation</name>
    <version>1.14.2</version>
    <description>
        A 2D navigation stack that takes in information from odometry, sensor
    </description>
    <maintainer email="linux.kbp@gmail.com">Kumar Bipin</maintainer>
    <author>linux.kbp@gmail.com</author>
    <license>BSD,LGPL,LGPL (amcl)</license>
    <url>http://wwww.kumar-bipin/navigation</url>

    <buildtool_depend>catkin</buildtool_depend>

    <run_depend>amcl</run_depend>
    <run_depend>carrot_planner</run_depend>
    .
    .
    .
    <run_depend>move_slow_and_clear</run_depend>
    <run_depend>voxel_grid</run_depend>

    <export>
        <metapackage/>
    </export>
</package>
```

Screenshot of package.xml

ROS uses a simplified message description language to describe the data values that ROS nodes publish. These datatype descriptions can be used to generate source code for the appropriate message type in different target languages such as C++, Python, Java, MATLAB, and so on.

The data type description of ROS messages is stored in the `.msg` files in the `msg` subdirectory of a ROS package. The message definition can consist of two types—fields and constants. The field is split into field types and field name. The field type is the data type of the transmitting message and the field name is its name. The constants define a constant value in the message file.

An example of an `.msg` file is as follows:

```
int32 id
float32 speed
string name
```

In ROS, we can find a set of standard types to use in messages, as shown in the following table:

Primitive type	Serialization	C++	Python
bool(1)	unsigned 8-bit int	uint8_t	bool
int8	signed 8-bit int	int8_t	int
uint8	unsigned 8-bit int	uint8_t	int(3)
int16	signed 16-bit int	int16_t	int
uint16	unsigned 16-bit int	uint16_t	int
int32	signed 32-bit int	int32_t	int
uint32	unsigned 32-bit int	uint32_t	int
int64	signed 64-bit int	int64_t	long
uint64	unsigned 64-bit int	uint64_t	long
float32	32-bit IEEE float	float	float
float64	64-bit IEEE float	double	float
string	ascii string(4)	std::string	string
time	secs/nsecs unsigned 32-bit ints	ros::Time	rospy.Time
duration	secs/nsecs signed 32-bit ints	ros::Duration	rospy.Duration

A special type of ROS message is message headers, which can carry information such as a time or stamp, a frame of reference or frame_id, and a sequence number or seq. Using this information, ROS messages can be transparent to the ROS nodes that are handling them.

The rosmsg command tool can be used to get the information about the message header:

```
$ rosmsg show std_msgs/Header
```

We can get the following outputs:

```
uint32 seq time stamp string frame_id
```

In the upcoming sections, we will look into how to create messages with the right tools.

ROS uses a simplified service description language to describe ROS service types. This is built directly upon the ROS msg format to enable request-response communication between nodes. Service descriptions are stored in the `.srv` files in the `srv` subdirectory of a package.

An example service description format is as follows:

```
#Request message type string req --- #Response message type string res
```

The first section is the type of requested message that is separated by `---` and the next section is the type of response message. In these examples, both `Request` and `Response` are strings.

In the upcoming sections, we will look into how to work with ROS services.

Analyzing the ROS computation graph

ROS creates a computational network where all the processes are connected. Any process in the system can access this network, interact with other processes, observe the information that they are sending, and transmit data to the network. In general, they perform collaborative computational tasks.

Getting ready

This computation network can be called the computation graph. The basic concepts in the computation graph are ROS nodes, master, parameter server, messages, topics, services, and bags. Each concept in the computational graph has a specific contribution in different ways.

The ROS communication-related packages, including core client libraries, such as `roscpp` and `rospython`, and the implementation of concepts such as topics, nodes, parameters, and services, are included in a metapackage called `ros_comm`. Moreover, this stack also consists of tools such as `rostopic`, `rosparam`, `rosservice`, and `rosnode`, which introspect the preceding concepts.

The `ros_comm` stack contains the ROS communication middleware packages, and these packages are collectively called the ROS graph layer, which is show in the following figure:

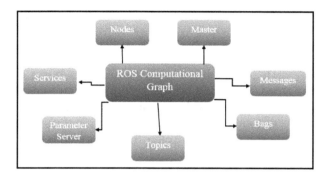

ROS computational graph

The following are abstracts of each graph's concepts:

- **ROS nodes**: ROS nodes correspond to the process in the Linux system that performs specific tasks. Also, ROS nodes are written using ROS client libraries such as `roscpp` and `rospy`, which support the implementation of different types of communication methods, mostly ROS messages and ROS services. Using the ROS communication methods, they can exchange data and create an ROS computational network. However, ROS has a design philosophy of having several nodes that provide only a single functionality, rather than having a large node that makes everything in the system. For example, in a robotic system, there will be different nodes to perform specific kinds of tasks such as perception, planning, and control.

- **The ROS master**: The ROS master provides the registration of names and the lookup service for the rest of the nodes and is responsible for setting up connections between the nodes. In an ROS system, nodes will not be able to find each other, exchange messages, or invoke services without a ROS master. However, in a distributed system, it could be possible to run the master on one computer, and other remote nodes can find each other by communicating with this master.

- **The ROS parameter server**: The parameter server allows the ROS system to keep the data or configuration information stored in a central place. All nodes can access and modify these values. The parameter server is a part of the ROS master.

- **ROS messages**: ROS nodes communicate with each other through messages, which contain data that provides information to other nodes. Although ROS has standard primitive types (integer, floating point, Boolean, and so on), it provides a mechanism to develop custom types of messages using standard message types.

- **ROS topics**: Each message in ROS is transported using named buses called topics. When a node sends a message through a topic, in ROS terminology, it is said that the node is publishing a topic. Similarly, when a node receives a message through a topic, it is said that the node is subscribing to a topic. However, as the publishing node and subscribing node are not aware of each other's existence, a node could even subscribe to a topic that might not have a publisher. This allows us to decouple production from consumption. Moreover, it's important that topic names are unique to avoid problems and confusion between topics with the same name.

- **ROS services**: In some robot applications, a publish-subscribe model will not be enough although it needs a request-response interaction. The publish-subscribe model is a kind of one-way transport system whereas when, working with a distributed system, a request-response kind of interaction might be required. ROS services are used in such cases. Also, the ROS service definition contains two parts – one is for requests and the other is for responses. In addition, it also requires two node servers and clients. The server node provides the service under a name, and when the client node sends a request message to this server, it will respond and send the result to the client. The client might need to wait until the server responds. The ROS service interaction is like a **remote procedure call** (**RPC**).

- **ROS bags**: ROS bags are a file format for saving and playing back ROS message data. These are an important mechanism for storing data, such as sensor data, which can be difficult to collect but is necessary for developing and testing robotic algorithms. Bags are very useful features for development and debugging. We will discuss ROS bags in upcoming chapters.

The following graph shows how the nodes communicate with each other using topics. The topics are represented in a rectangle and nodes are represented in ellipses. The messages and parameters are not included in this graph. The tool used to create the graph is rqt_graph; we will discuss it in detail in Chapter 3, *ROS Visualization and Debugging Tools*:

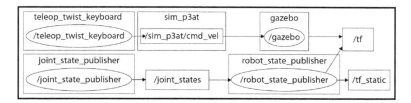

Communication graph

How it works...

In the previous section, we learned about important elements in the ROS computational network to start with. Following that, in this section, we will warm up using these elements. Let's get started with practicing with the ROS master.

ROS master works much like a DNS server. Whenever any node starts in the ROS system, it will start looking for ROS master and register the name of the node with ROS master. Therefore, ROS master has information about all the nodes that are currently running on the ROS system. When information about any node changes, it will generate a call back and update with the latest information.

The node will provide the details of the topic, such as its name and data type, to ROS master before the node starts publishing that topic. ROS master will check whether any other nodes are subscribed to the same topic. If any nodes are subscribed to the same topic, ROS master will share the node details of the publisher to the subscriber node. After receiving the node details, these two nodes will interconnect using the TCPROS protocol, which is based on TCP/IP sockets, and ROS master will relinquish its role in controlling them. In the future, we might stop either the publisher node or the subscriber node, according to our requirements. Whenever the state of a node changes, it will check with ROS master once again. The ROS service also uses the same method. The nodes are written using the ROS client libraries such as `roscpp` and `rospy`. These clients interact with ROS master using **XML Remote Procedure Call** (**XMLRPC**) based APIs, which act as the backend of the ROS system APIs.

However, in a distributed network, in which different physical computers participate in a ROS computational network, ROS_MASTER_URI should be defined properly. Thereupon, the remote nodes can find each other and communicate with each other. A ROS system must have only one master, even in a distributed system, and it should run on a computer that is reachable by all other computers to ensure that remote ROS nodes can access the master.

The following diagram shows an illustration of how ROS master interacts with a publishing and subscribing node:

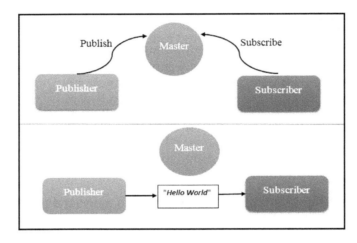

Communication setup for publisher and subscriber nodes with ROS master

Now, let's look at practicing with ROS nodes.

ROS nodes are processes that perform computation using ROS client libraries such as roscpp and rospy. One node can communicate with other nodes using ROS topics and services and a parameters server.

Moreover, using nodes in ROS provides fault tolerance and separates the code and functionalities, making the system simpler and robust. A robotic application might contain several nodes, where one node processes camera images, the next one handles serial data from the robot, another one can be used to compute odometry, and so on. Also, nodes reduce complexity and increase debugability compared to monolithic codes.

There is a `rosbash` tool for introspecting ROS nodes. The `rosnode` tool is a command-line tool used to display information about nodes, such as listing the currently running nodes.

Following are the explanations of the commands:

- `rosnode info NODE`: Prints information about a node
- `rosnode kill NODE`: Kills a running node or sends a given signal
- `rosnode list`: Lists the active nodes
- `rosnode machine hostname`: Lists the nodes running on a particular machine or lists machines
- `rosnode ping NODE`: Tests the connectivity to the node
- `rosnode cleanup`: Purges the registration information from unreachable nodes

In upcoming sections, we will discuss the working of ROS nodes that use functionalities such as ROS topics, services, and messages.

Let's look at practicing with ROS topics.

ROS topics are named buses through which ROS nodes exchange messages that can anonymously publish and subscribe, which means that the production of messages is decoupled from the consumption. In fact, the ROS nodes are not interested in knowing which node is publishing the topic or subscribing to topics, they only look for the topic name and whether the message types of the publisher and subscriber match.

A topic can have various subscribers and can also have various publishers, but we should be careful when publishing the same topic with different nodes as it can create conflicts. However, each topic is strongly typed by the ROS message type used to publish it, and nodes can only receive messages from a matching type. A node can subscribe to a topic only if it has the same message type.

The ROS nodes communicate with topics using TCP/IP-based transport known as TCPROS. This method is the default transport method used in ROS. Another type of communication is UDPROS, which has low-latency, loose transport, and is only suited for teleoperation.

ROS has a tool to work with topics called `rostopic`. It is a command-line tool that gives us information about the topic or publishes data directly on the ROS computational network.

Following are the explanations of the commands:

- `rostopic bw /topic`: Displays the bandwidth used by the topic
- `rostopic echo /topic`: Prints messages to the screen
- `rostopic find message_type`: Finds topics by their type
- `rostopic hz /topic`: Displays the publishing rate of the topic
- `rostopic info /topic`: Prints information about the topic, such as its message type, publishers, and subscribers
- `rostopic list`: Prints information about active topics
- `rostopic pub /topic type args`: Used to publish a value to a topic with a message type
- `rostopic type /topic`: Displays the message type of the given topic

We will learn how to use this command-line tool in upcoming sections.

For practicing with ROS messages, see the following:

- ROS nodes communicate with each other by publishing messages to a topic. As discussed earlier, messages are simple data structures that use standard types or custom types developed by the user.
- Message types use the following standard ROS naming convention: the name of the package, then / and then the name of the `.msg` file. For example, `std_msgs/msg/String.msg` has the `std_msgs/String` message type. In addition to message data type, ROS uses an MD5 checksum comparison to confirm whether the publisher and subscriber exchange the same message data types.

- ROS has inbuilt tools called `rosmsg` for getting information about ROS messages.

  Following are the explanations of the commands:

  - `rosmsg show [message]`: Shows the message description
  - `rosmsg list`: Lists all messages
  - `rosmsg md5 [message]`: Displays MD5 checksum of a message
  - `rosmsg package [package_name]`: Lists messages in a package
  - `rosmsg packages [package_1] [package_2]`: Lists packages that contain messages

Let's see practicing with ROS services:

- When we need a request-response kind of communication in ROS, we have to use ROS services. ROS topics can't do this kind of communication because it is unidirectional. ROS services are mainly used in a distributed system.
- The ROS service is defined using a pair of messages, a request datatype, and a response datatype in a `srv` file. The `srv` files are kept in an `srv` folder inside a package.
- In ROS services, one server node acts as a ROS server, in which the client nodes can request for a specific service. If the server completes the service request, it will send the results to the client.
- Similar to topics, services have an associated service type that is the package resource name of the `.srv` file. As with other ROS filesystem-based types, the service type is the package name and the name of the `.srv` file. For example, the `chapter2_tutorials/srv/chapter2_srv.srv` file has the `chapter2_tutorials/chapter2_srv` service type. In ROS services also, there is an MD5 checksum that checks in the nodes. If the sum is equal, then only the server responds to the client.
- ROS has two command-line tools to work with services, `rossrv` and `rosservice`. The first tool is `rossrv`, which is similar to `rosmsg`, and is used to get information about service types. The next command is `rosservice`, which is used to list and query the running ROS services.

Following are the explanations of the commands:

- `rosservice call /service args`: Calls the service using the given arguments
- `rosservice find service_type`: Finds services in the given service type
- `rosservice info /services`: Prints information about the given service
- `rosservice list`: Lists the active services running on the system
- `rosservice type /service`: Prints the service type of a given service
- `rosservice uri /service`: Prints the service ROSRPC URI

Now, let's move on to the next one, practicing with the ROS parameter server:

- The parameter server is a shared, multivariable dictionary that is accessible through the ROS computational network. Nodes use this server to store and retrieve parameters at runtime.
- A parameter server is implemented using XMLRPC and runs inside the ROS master, which means that its API is accessible through normal XMLRPC libraries. XMLRPC is an RPC protocol that uses XML to encode its calls and HTTP as a transport mechanism.

The parameter server supports the following XMLRPC datatypes:

- 32-bit integers
- Booleans
- Strings
- Doubles
- ISO 8601 dates
- Lists
- Base64-encoded binary data

ROS has the `rosparam` tool for working with parameter servers. The parameters can be changed dynamically during the execution of the node that uses these parameters by using the `dyamic_reconfigure` package. We will discuss this further in an upcoming section.

Following are the explanations of the commands:

- `rosparam set [parameter_name] [value]`: Sets a value in the given parameter
- `rosparam get [parameter_name]`: Retrieves a value from the given parameter
- `rosparam load [YAML file]`: Loads the parameter from a saved YAML file
- `rosparam dump [YAML file]`: Dumps the existing ROS parameters to a YAML file
- `rosparam delete [parameter_name]`: Deletes the given parameter
- `rosparam list`: Lists existing parameter names

Let's practice with ROS bags:

- A bag is a file created by ROS with the `.bag` format to save all of the information of the messages, topics, services, and others. We can use this data later to visualize what has happened; we can play, stop, rewind, and perform other operations with it.
- Bag files are created using the `rosbag` command. The main application of `rosbag` is data logging. The logged ROS computational data can be visualized and processed offline.

Following are the explanations of the commands:

- `rosbag record [topic_1] [topic_2] -o [bag_name]`: Records the given topics into a bag file that is given in the command
- `rosbag -a [bag_name]`: Records all
- `rosbag play [bag_name]`: Plays back the existing bag file

Associating with the ROS community

ROS is a large project that has many ancestors and contributors. The need for an open collaboration framework was felt by many people in the robotics research community. The effort was boosted by countless researchers who contributed their time and expertise to the core of ROS and its fundamental software packages. Throughout, the software was developed in the open using the permissive BSD open source license, and it gradually became widely used in the robotics research community.

The ROS ecosystem now consists of tens of thousands of users worldwide, working in domains ranging from tabletop hobby projects to large industrial automation systems.

Getting ready

The ROS community-level concepts are the ROS resources that enable separate communities to exchange software and knowledge. These resources include the following:

- **ROS distributions**: ROS distributions are collections of versioned metapackages that can be installed. ROS distributions play a similar role to Linux distributions. They enable easier installation and collection of the ROS software. ROS distributions maintain consistent versions across a set of software.
- **ROS repositories**: ROS relies on a federated network of code repositories, where different institutions can develop and release their own robot software components.
- **The ROS wiki**: The ROS community wiki is the main forum for documenting information about ROS. Anyone can sign up for an account and contribute their own documentation, provide corrections or updates, write tutorials, and more.
- **ROS bug ticket system**: If we find a bug in the existing software or need to add a new feature, we can use this resource.
- **ROS mailing lists**: The ROS user-mailing list is the primary communication channel about new updates to ROS, as well as a forum for asking questions about the ROS software.
- **ROS answers**: This website resource helps to ask questions related to ROS. If we post our doubts on this site, other ROS users can see this and suggest solutions.
- **ROS blog**: The ROS blog updates with news, photos, and videos related to the ROS community (`http://www.ros.org/news`).

Learning working with ROS

In the preceding section, we learned about the basic concept and terminology required to start with the ROS computational network framework. Subsequently, in this section, we will focus mostly on the practical aspect.

Getting ready

Before running any ROS nodes, we should start the ROS master and the ROS parameter server. We can start ROS master and the ROS parameter server using a single command called `roscore`, which will start three programs by default; ROS master, ROS parameter server, and `rosout` logging nodes.

How to do it...

It is time for us to practice what we have learned until now. Following, we will see examples to practice, along with creating packages, using nodes, using parameter servers, and moving a simulated robot with `turtlesim`.

To run ROS master and parameter server, use the following command:

```
$ roscore
```

In the following screenshot, we can see a log file is created inside the `/home/kbipin/.ros/log` folder for collecting logs from ROS nodes, in the first part of the screen. This file can be used for debugging purposes.

The next part shows that the `theroslaunch` command is executing a ROS launch file called `roscore.xml`. When a launch file executes, it automatically starts `rosmaster` and the ROS parameter server. The `roslaunch` command is a Python script, which can start `rosmaster` and the ROS parameter server whenever it tries to execute a launch file.

It also shows the address of the ROS parameter server within the port:

```
kbipin@ubuntu:~/catkin_ws/devel$ roscore
... logging to /home/kbipin/.ros/log/00cf43da-6acd-11e7-a3fd-000c29ba2ae0/roslaunch-ubuntu-2313.log
Checking log directory for disk usage. This may take awhile.
Press Ctrl-C to interrupt
Done checking log file disk usage. Usage is <1GB.

started roslaunch server http://ubuntu:36055/
ros_comm version 1.12.7

SUMMARY
========

PARAMETERS
 * /rosdistro: kinetic
 * /rosversion: 1.12.7

NODES

auto-starting new master
process[master]: started with pid [2324]
ROS_MASTER_URI=http://ubuntu:11311/

setting /run_id to 00cf43da-6acd-11e7-a3fd-000c29ba2ae0
process[rosout-1]: started with pid [2337]
started core service [/rosout]
```

Terminal messages while running the roscore command

In the third part of the screen, we can see parameters such as rosdistro and rosversion displayed on the terminal. These parameters are displayed when the roslaunch command executes roscore.xml.

In the fourth section, we can see that the rosmaster node is started using ROS_MASTER_URI, which is an environment variable. As discussed in the previous section, it provides the address. In the last part, we can see that the rosout node is started, which will start subscribing the /rosout topic and rebroadcasting into /rosout_agg.

When the roscore command is executed, initially, it checks the command line argument for a new port number for rosmaster. If it gets the port number, it will start listening to the new port number, otherwise it will use the default port. This port number and the roscore.xml launch file will pass to the roslaunch system. The roslaunch system is implemented in a Python module; it will parse the port number and launch the roscore.xml file.

The content of the roscore.xml file is as follows:

```
<launch>
  <group ns="/">
    <param name="rosversion" command="rosversion roslaunch" />
    <param name="rosdistro" command="rosversion -d" />
    <node pkg="rosout" type="rosout" name="rosout" respawn="true"/>
```

```
    </group>
  </launch>
```

In the `roscore.xml` file, we can see the ROS parameters and nodes are encapsulated in a group XML tag with a / namespace. The group XML tag indicates that all the nodes inside this tag have the same settings. The two parameters, called `rosversion` and `rosdistro`, store the output of the `rosversion roslaunch` and `rosversion -d` commands using the command tag, which is a part of the ROS `param` tag. The `command` tag will execute the command mentioned on it and store the output of the command in these two parameters.

The `rosout` node will collect log messages from other ROS nodes and store them in a log file, and will also rebroadcast the collected log messages to another topic. The topic `/rosout` is published by ROS nodes working using ROS client libraries such as `roscpp` and `rospy` and this topic is subscribed by the `rosout` node, which rebroadcasts the message in another topic called `/rosout_agg`. This topic has an aggregate stream of log messages.

Let's check the ROS topics and ROS parameters created after running `roscore`. The following command will list the active topics on the Terminal:

```
$ rostopic list
```

The list of topics is as follows:

```
/rosout
/rosout_agg
```

The following command lists out the parameters available when running `roscore`. The following is the command to list the active ROS parameter:

```
$ rosparam list
```

The parameters are mentioned here; they have the ROS distribution name, version, and address of `roslaunch` server and run_id, where run_id is a unique ID associated with a particular run of `roscore`:

```
/rosdistro
/roslaunch/uris/host_ubuntu__33187
/rosversion
/run_id
```

A list of the ROS services generated during the running of `roscore` can be checked using the following command:

```
$ rosservice list
```

The list of services running is as follows:

```
/rosout/get_loggers
/rosout/set_logger_level
```

These ROS services are generated for each ROS node for setting the logging levels. After understanding the basics of ROS master, the parameter server, and roscore, we can revisit the concepts of ROS nodes, topics, messages, and services in more detail.

As discussed in the previous section, nodes are executable programs, and, once built, these executables can be found in the devel space. To practice and learn about nodes, we will use a typical package called turtlesim.

The turtlesim package is preinstalled with the desktop installation by default; if not, we can install it with the following command:

```
$ sudo apt-get install ros-kinetic-ros-tutorials
```

In the previous section, we have started the roscore in one of our open terminals.

Now, we are going to start a new node in another Terminal with rosrun, using the following command:

```
$ rosrun turtlesim turtlesim_node
```

We will then see a new window appear with a little turtle in the middle, as shown in the following screenshot:

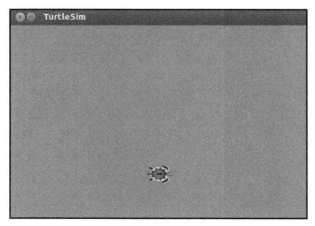

Turtlesim

We are going to get information about the nodes that are running, using the following command:

```
$ rosnode list
```

We will see a new node with the name /turtlesim. We can get information about the node using the following command:

```
$ rosnode info /turtlesim
```

The previous command will print the following information:

```
Node [/turtlesim]
Publications:
 * /turtle1/color_sensor [turtlesim/Color]
 * /rosout [rosgraph_msgs/Log]
 * /turtle1/pose [turtlesim/Pose]

Subscriptions:
 * /turtle1/cmd_vel [unknown type]

Services:
 * /turtle1/teleport_absolute
 * /turtlesim/get_loggers
 * /turtlesim/set_logger_level
 * /reset
 * /spawn
 * /clear
 * /turtle1/set_pen
 * /turtle1/teleport_relative
 * /kill


contacting node http://ubuntu:37703/ ...
Pid: 2422
Connections:
 * topic: /rosout
    * to: /rosout
    * direction: outbound
    * transport: TCPROS
```

Node information

In the information, we can see the publications (topics), subscriptions (topics), and services (srv) that the node has and the unique name of each.

In the next section, we will learn how to interact with the node using topics and services.

We have the rostopic tool to interact with and get information about topics. Using the rostopic pub, we can publish topics that can be subscribed to by any node. We have to publish the topic with the correct name.

We will start the `turtle_teleop_key` node in the `turtlesim` package with the following command:

```
$ rosrun turtlesim turtle_teleop_key
```

With this node, we can move the turtle using the arrow keys, as shown in the following screenshot:

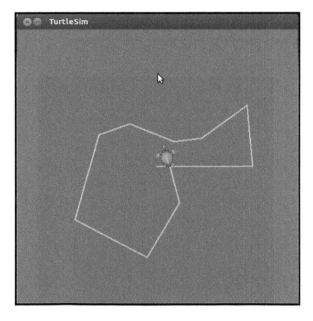

Turtlesim teleoperation

Let us understand why the turtle moves when `turtle_teleop_key` is running. Looking into the information provided by the `rosnode` information about the `/teleop_turtle` and `/turtlesim` nodes, we can notice that there exists a topic called `/turtle1/cmd_vel` [geometry_msgs/Twist] in the publications section of the `/teleop_turtle` node, and in the subscriptions section of the `/turtlesim` node, there is `/turtle1/cmd_vel` [geometry_msgs/Twist]:

```
$ rosnode info /teleop_turtle
```

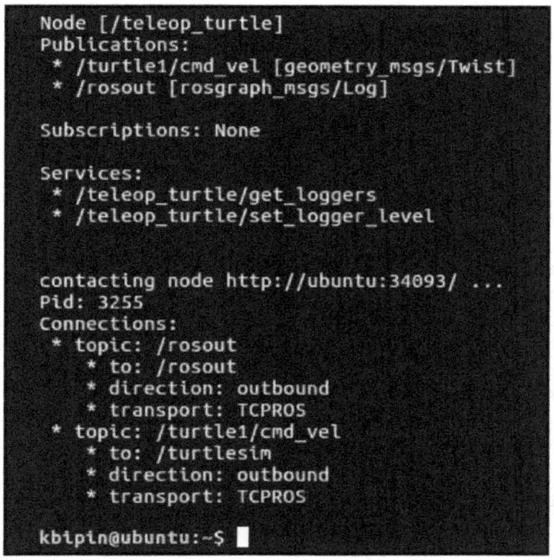

```
Node [/teleop_turtle]
Publications:
 * /turtle1/cmd_vel [geometry_msgs/Twist]
 * /rosout [rosgraph_msgs/Log]

Subscriptions: None

Services:
 * /teleop_turtle/get_loggers
 * /teleop_turtle/set_logger_level

contacting node http://ubuntu:34093/ ...
Pid: 3255
Connections:
 * topic: /rosout
    * to: /rosout
    * direction: outbound
    * transport: TCPROS
 * topic: /turtle1/cmd_vel
    * to: /turtlesim
    * direction: outbound
    * transport: TCPROS

kbipin@ubuntu:~$ 
```

Teleop node information

This shows that the first node is publishing a topic that the second node can subscribe to. We can observe the topic list using the following command line:

```
$ rostopic list
```

The output will be as follows:

```
/rosout
/rosout_agg
/turtle1/colour_sensor
/turtle1/cmd_vel
/turtle1/pose
```

We can get the information sent by the node using the `echo` parameter by running the following command:

```
$ rostopic echo /turtle1/cmd_vel
```

We will see something similar to the following output:

```
---
linear:
x: 0.0
y: 0.0
z: 0.0
angular:
x: 0.0
```

```
y: 0.0
z: 2.0
---
```

Similarly, we can get the type of message sent by `topic` using the following command line:

```
$ rostopic type /turtle1/cmd_vel
```

The output will be as follows:

```
Geometry_msgs/Twist
```

For getting the message fields, we can use the following command:

```
$ rosmsg show geometry_msgs/Twist
```

We will get something similar to the following output:

```
geometry_msgs/Vector3 linear
  float64 x
  float64 y
  float64 z
geometry_msgs/Vector3 angular
  float64 x
  float64 y
  float64 z
```

This is useful information. We can publish topics using the `rostopic pub [topic]` `[msg_type] [args]` command:

```
$ rostopic pub /turtle1/cmd_vel  geometry_msgs/Twist -r 1 --
"linear:
x: 1.0
y: 0.0
z: 0.0
angular:
x: 0.0
y: 0.0
z: 1.0"
```

We can observe the turtle doing a curve, as shown in the following screenshot:

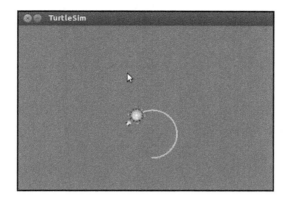

Turtle doing a curve

Services are another method through which nodes can communicate with each other, and which allow nodes to send a request and receive a response. As discussed in the preceding section, *Understating ROS services*, we have the rosservice tool to interact with and get information about services.

The following command will list the services available for the turtlesim node (run roscore and the turtlesim node if they are not currently running):

```
$ rosservice list
```

The following output will be shown:

```
/clear
/kill
/reset
/rosout/get_loggers
/rosout/set_logger_level
/spawn
/teleop_turtle/get_loggers
/teleop_turtle/set_logger_level
/turtle1/set_pen
/turtle1/teleport_absolute
/turtle1/teleport_relative
/turtlesim/get_loggers
/turtlesim/set_logger_level
```

The following command can be used to get the type of any service—for example, the /clear service:

```
$ rosservice type /clear
```

The following output will be shown:

```
std_srvs/Empty
```

To invoke a service, we can use rosservice call [service] [args]. For example, the following command will invoke the /clear service:

```
$ rosservice call /clear
```

In the turtlesim window, we can now see that the lines created by the movements of the turtle will be deleted.

Now, we can look into another service, for example, the /spawn service. This will create another turtle at a given location and with a specified orientation. Let's start with the following command:

```
$ rosservice type /spawn | rossrv show
```

The output will be as follows:

```
float32 x
float32 y
float32 theta
string name
---
string name
```

The preceding command is the same as the following commands (for more detail, search in Google for *piping Linux*):

```
$ rosservice type /spawn
$ rossrv show turtlesim/Spawn
```

We will get something similar to the output shown by the previous command. We can know how to invoke the /spawn service by knowing its fields. It requires the position of x and y, the orientation (theta), and the name of the new turtle:

```
$ rosservice call /spawn 3 3 0.2 "new_turtle"
```

We will obtain the following output in the Turtlesim window:

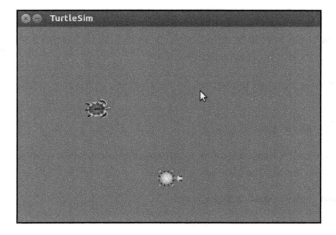

Turtlesim spawn

The ROS parameter server is used to store data at a central location that is accessible to all running nodes in a ROS computational network. As we have learned in the preceding section, *Understating ROS parameter server*, we have the `rosparam` tool to manage parameter servers.

For example, we can see the parameters in the server that are used by all currently running nodes:

```
$ rosparam list
```

We will obtain the following output:

```
/background_b
/background_g
/background_r
/rosdistro
/roslaunch/uris/host_ubuntu__33187
/rosversion
/run_id
```

The background parameters of the `turtlesim` node define the color of the windows, which is initially blue. We can read a value by using the `get` parameter:

```
$ rosparam get /background_b
```

Similarly, to set a new value, we can use the `set` parameter:

```
$ rosparam set /background_b 120
```

One of the most important features of `rosparam` is the `dump` parameter, which is used to save or load the contents of the parameter server.

To save the contents of a parameter server, we can use `rosparam` dump `[file_name]` as follows:

```
$ rosparam dump save.yaml
```

Similarly, to load a file with new data for a parameter server, we can use `rosparam load` `[file_name]` `[namespace]`, as follows:

```
$ rosparam load load.yaml namespace
```

How it works...

In the previous section, we learned about ROS master, topic, service, and parameter servers, with the example of the `turtlesim` package. This enables us to get ready for more advanced concepts of ROS. In this section, we will learn about creating and building example nodes, as well as about the `msg` and `srv` files.

Creating ROS nodes

In this section, we are going to learn how to create two nodes, where one will publish data and the other will receive this data. This is the most basic way of communicating between two nodes in the ROS system.

We created the ROS package *chapter2_tutorials* in the earlier section *Creating a ROS package and metapackage*. Now, we will navigate to the `chapter2_tutorials/src/` folder using the following command:

```
$ roscd chapter2_tutorials/src/
```

We will create two files with the names `example_1a.cpp` and `example_1b.cpp`. The `example_1a.cpp` file will create a node named `example1a`, which publishes the data "Hello World!" on the `/message` topic. Moreover, the `example_1b.cpp` file will create a node named `example1b`, which subscribes to the `/message` topic and receives the data sent by the `example1a` node, and shows the data in the shell.

We can copy the following code inside the `example_1a.cpp` file, or download it from the repository (`https://github.com/kbipin/Robot-Operating-System-Cookbook/blob/master/chapter2_tutorials/src/example_1a.cpp`):

```
#include "ros/ros.h"
#include "std_msgs/String.h"
#include <sstream>

int main(int argc, char **argv)
{
  ros::init(argc, argv, "example1a");
  ros::NodeHandle n;
  ros::Publisher pub = n.advertise<std_msgs::String>("message", 100);
  ros::Rate loop_rate(10);
  while (ros::ok())
  {
    std_msgs::String msg;
    std::stringstream ss;
    ss << "Hello World!";
    msg.data = ss.str();
    pub.publish(msg);
    ros::spinOnce();
    loop_rate.sleep();
  }
  return 0;
}
```

We will discuss the preceding code to understand the ROS development framework in detail. The headers to be included are `ros/ros.h`, `std_msgs/String.h`, and `sstream`. `ros/ros.h` includes all the files needed for working with ROS, and `std_msgs/String.h` includes the header that defines the type of message used for publishing to the ROS computational network:

```
#include "ros/ros.h"
#include "std_msgs/String.h"
#include <sstream>
```

Here, we will initialize the node and set the name. It should be remembered that the name must be unique:

```
ros::init(argc, argv, "example1a");
```

This is the handler of the processes associated with the node, which allow it to interact with the environment:

```
ros::NodeHandle n;
```

At this point, we will instantiate a publisher with the name of the topic and the type, which also specifies the size of a buffer in the second parameter (if the topic is publishing data quickly, the buffer will keep at least 100 messages):

```
ros::Publisher pub = n.advertise<std_msgs::String>("message", 100);
```

The next line of code sets the data sending frequency, which, in this case, is 10 Hz:

```
ros::Rate loop_rate(10);
```

The next `ros::ok()` line stops the node if *Ctrl* + *C* is pressed or if ROS stops all the nodes:

```
while (ros::ok())
{
```

In this part of the code, we will create a variable for the message with the correct type to send the data:

```
std_msgs::String msg;
std::stringstream ss;
ss << "Hello World!";
msg.data = ss.str();
pub.publish(msg);
```

Moreover, we will continue by sending the message, using the previously defined semantics of the publisher:

```
pub.publish(msg);
```

The `spinOnce` function takes care of handling all of the internal ROS events and actions, such as reading from subscribed topics; however, `spinOnce` performs one iteration in the main loop of ROS, in order to allow the user to perform actions between iterations, while the `spin` function runs the main loop without interruption.

Finally, at this point, we sleep for the required time to get a 10 Hz frequency:

```
loop_rate.sleep();
```

We have successfully created a publisher node. Likewise, we will now create the subscriber node. Copy the following code inside the `example_1b.cpp` file, or download it from the repository (https://github.com/kbipin/Robot-Operating-System-Cookbook/blob/master/chapter2 _tutorials/src/example_1b.cpp):

```cpp
#include "ros/ros.h"
#include "std_msgs/String.h"

void messageCallback(const std_msgs::String::ConstPtr& msg)
{
  ROS_INFO("Thanks: [%s]", msg->data.c_str());
}

int main(int argc, char **argv)
{
  ros::init(argc, argv, "example1b");
  ros::NodeHandle n;
  ros::Subscriber sub = n.subscribe("message", 100, messageCallback);
  ros::spin();
  return 0;
}
```

Let's discuss the source code. As discussed previously, `ros/ros.h` includes all the files necessary for using the node with ROS, and `std_msgs/String.h` defines the type of message used:

```cpp
#include "ros/ros.h"
#include "std_msgs/String.h"
#include <sstream>
```

The following source code shows the type of function callback that is invoked in response to an action, which in this case is the reception of a string message on subscribed topics. This function allows us to process the received message data; in this case, it displays the message data on the Terminal:

```cpp
void messageCallback(const std_msgs::String::ConstPtr& msg)
{
  ROS_INFO("Thanks: [%s]", msg->data.c_str());
}
```

At this point, we will create a subscriber and start to listen to the topic with the name message where the buffer size will be of 100, and the function to handle the message will be `messageCallback`:

```
ros::Subscriber sub = n.subscribe("message", 1000, messageCallback);
```

Finally, the line `ros::spin()` is the main loop where the node begins to read the topics, and, when a message arrives, `messageCallback` is called. When the user presses *Ctrl + C*, the node exits the loop and ends:

```
ros::spin();
```

Building the ROS node

As we are working with the `chapter2_tutorials` package, we will edit the `CMakeLists.txt` file here and prepare and configure the packages for the build:

```
$ rosed chapter2_tutorials CMakeLists.txt
```

At the end of this file, we will have to add the following lines (https://github.com/kbipin/Robot-Operating-System-Cookbook/blob/master/chapter2_tutorials/CMakeLists.txt):

```
include_directories(
include
  ${catkin_INCLUDE_DIRS}
)

add_executable(example1a src/example_1a.cpp)
add_executable(example1b src/example_1b.cpp)

add_dependencies(example1a chapter2_tutorials_generate_messages_cpp)
add_dependencies(example1b chapter2_tutorials_generate_messages_cpp)

target_link_libraries(example1a ${catkin_LIBRARIES})
target_link_libraries(example1b ${catkin_LIBRARIES})
```

The `catkin_make` tool is used to build the package that will compile all the nodes:

```
$ cd ~/catkin_ws/
$ catkin_make --pkg chapter2_tutorials
```

To work with the nodes created previously, we have to first start `roscore`:

```
$ roscore
```

Next, we can check whether ROS is running using the `rosnode list` command in another terminal, as follows:

```
$ rosnode list
```

Now, we run both nodes in different shells:

```
$ rosrun chapter2_tutorials example1a
$ rosrun chapter2_tutorials example1b
```

We will see something similar to the following screenshot in the shell where the `example1b` node is running:

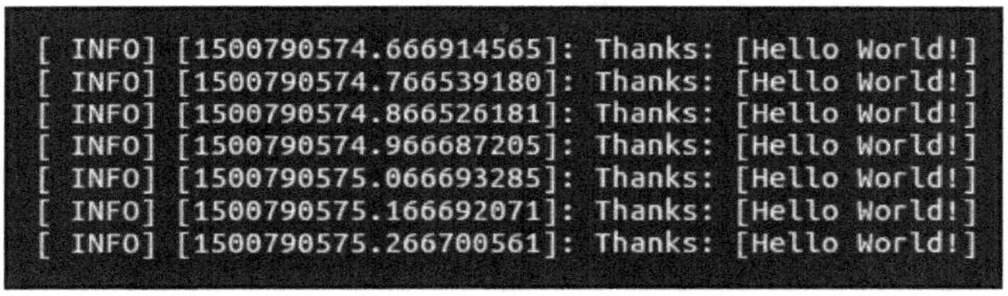

```
[ INFO] [1500790574.666914565]: Thanks: [Hello World!]
[ INFO] [1500790574.766539180]: Thanks: [Hello World!]
[ INFO] [1500790574.866526181]: Thanks: [Hello World!]
[ INFO] [1500790574.966687205]: Thanks: [Hello World!]
[ INFO] [1500790575.066693285]: Thanks: [Hello World!]
[ INFO] [1500790575.166692071]: Thanks: [Hello World!]
[ INFO] [1500790575.266700561]: Thanks: [Hello World!]
```

Subscriber node output

We can use the `rosnode` and `rostopic` commands to debug and get the information about the running node:

```
$ rosnode list
$ rosnode info /example1_a
$ rosnode info /example1_b
$ rostopic list
$ rostopic info /message
$ rostopic type /message
$ rostopic bw /message
```

Creating ROS messages

In this section, we will learn how to create a user-defined custom message using the `.msg` file, which will be used in our nodes. This contains a specification about the type of data to be transmitted. The ROS build system will use this file to create the necessary code to implement the message in the ROS computational framework or network.

In the previous section, *Creating nodes*, we created two nodes with a standard type message. Now, we will learn how to create custom messages with ROS tools.

First, create a `msg` folder in the `chapter2_tutorials` package. Furthermore, create a new `chapter2_msg.msg` file there and add the following lines:

```
int32 A
int32 B
int32 C
```

Furthermore, we will have to search for the following lines and uncomment them in the `package.xml` file:

```
<build_depend>message_generation</build_depend>
<run_depend>message_runtime</run_depend>
```

These lines enable the configuration of the messages and services in the ROS build system. Also, we will add the `message_generation` line in `CMakeList.txt`:

```
find_package(catkin REQUIRED COMPONENTS
  roscpp
  std_msgs
  message_generation
)
```

Additionally, we have to find and uncomment the `add_message_files` lines in `CMakeList.txt`, along with adding the name of the new message file, as follows:

```
## Generate messages in the 'msg' folder
add_message_files(
        FILES
        chapter2_msg.msg
)

## Generate added messages and services with any dependencies listed here
 generate_messages(
   DEPENDENCIES
   std_msgs
 )
```

Finally, we can build the package using the following command:

```
$ cd ~/catkin_ws/
$ catkin_make
```

To check whether everything is OK, we can use the `rosmsg` command:

```
$ rosmsg show chapter2_tutorials/chapter2_msg
```

This will show the contents of the `chapter2_msg.msg` file.

At this moment, we will create nodes that use the custom msg file described previously. This will be very similar to `example_1a.cpp` and `example_1b.cpp`, as discussed previously, but with the new message, `chapter2_msg.msg`.

The following code snippet is present in the `example_2a.cpp` file (`https://github.com/kbipin/Robot-Operating-System-Cookbook/blob/master/chapter2_tutorials/src/example_2a.cpp`):

```cpp
#include "ros/ros.h"
#include "chapter2_tutorials/chapter2_msg.h"
#include <sstream>

int main(int argc, char **argv)
{
  ros::init(argc, argv, "example2a");
  ros::NodeHandle n;
  ros::Publisher pub =
n.advertise<chapter2_tutorials::chapter2_msg>("chapter2_tutorials/message",
100);
  ros::Rate loop_rate(10);
  while (ros::ok())
  {
    chapter2_tutorials::chapter2_msg msg;
    msg.A = 1;
    msg.B = 2;
    msg.C = 3;
    pub.publish(msg);
    ros::spinOnce();
    loop_rate.sleep();
  }
  return 0;
}
```

Similarly, the following code snippet is present in the `example_2b.cpp` file (`https://github.com/kbipin/Robot-Operating-System-Cookbook/blob/master/chapter2_tutorials/src/example_2b.cpp`):

```cpp
#include "ros/ros.h"
#include "chapter2_tutorials/chapter2_msg.h"
```

```
void messageCallback(const chapter2_tutorials::chapter2_msg::ConstPtr& msg)
{
  ROS_INFO("I have received: [%d] [%d] [%d]", msg->A, msg->B, msg->C);
}

int main(int argc, char **argv)
{
  ros::init(argc, argv, "example3_b");
  ros::NodeHandle n;
  ros::Subscriber sub = n.subscribe("chapter2_tutorials/message", 100,
messageCallback);
  ros::spin();
  return 0;
}
```

We can run both the publisher and the subscriber node with the following command:

```
$ rosrun chapter2_tutorials example2a
$ rosrun chapter2_tutorials example2b
```

While running both nodes in two separate shells, we will see something similar to the following output:

```
. . .
[ INFO] [1355280835.903686201]: I have received: [1] [2] [3]
[ INFO] [1355280836.020326872]: I have received: [1] [2] [3]
[ INFO] [1355280836.120367649]: I have received: [1] [2] [3]
[ INFO] [1355280836.220260466]: I have received: [1] [2] [3]
. . .
```

Creating ROS services

In this section, we will to learn how to create a .srv file, which will be used in our nodes. It contains a specification about the type of data to be transmitted. The ROS build system will use this file to create the necessary code to implement the srv in the ROS computational framework or network.

In the previous section, *Creating ROS messages,* we created two nodes with a custom type message. Now, we will learn how to create services with ROS tools.

First, create an `srv` folder in the `chapter2_tutorials` package. Furthermore, create a new `chapter2_srv.srv` file there and add the following lines:

```
int32 A
int32 B
---
int32 sum
```

Here, `A` and `B` are data types for a request from the client and `sum` is the response data type from the server.

We will have to search for the following lines and uncomment them:

```
<build_depend>message_generation</build_depend>
<exec_depend>message_runtime</exec_depend>
```

These lines enable the configuration of messages and services in the ROS build system. Also, we will add the `message_generation` line in `CMakeList.txt`:

```
find_package(catkin REQUIRED COMPONENTS
  roscpp
  std_msgs
  message_generation
)
```

Additionally, we will have to find and uncomment the `add_service_file` lines in `CMakeList.txt`, along with adding the name of the new message file, as follows:

```
## Generate services in the 'srv' folder
add_service_files(
    FILES
    chapter2_srv.srv
)
## Generate added messages and services with any dependencies listed here
  generate_messages(
    DEPENDENCIES
    std_msgs
  )
```

Finally, we can build the package using the following command:

```
$ cd ~/catkin_ws/
$ catkin_make
```

To check whether everything is OK, we can use the `rossrv` command:

```
$ rossrv show chapter2_tutorials/chapter2_srv
```

As of this moment, we have learned how to create the service data type in ROS. Following, we will look into the creation of a service that will calculate the sum of two numbers. We will create two nodes, a server and a client, in the `chapter2_tutorials` package, with the following names: `example_3a.cpp` and `example_3b.cpp`. Create them in the `src` folder.

In the first file, `example_3a.cpp`, we will the have following code (https://github.com/kbipin/Robot-Operating-System-Cookbook/blob/master/chapter2 _tutorials/src/example_3a.cpp):

```
#include "ros/ros.h"
#include "chapter2_tutorials/chapter2_srv.h"

bool add(chapter2_tutorials::chapter2_srv::Request  &req,
         chapter2_tutorials::chapter2_srv::Response &res)
{
  res.sum = req.A + req.B;
  ROS_INFO("Request: A=%d, B=%d", (int)req.A, (int)req.B);
  ROS_INFO("Response: [%d]", (int)res.sum);
  return true;
}

int main(int argc, char **argv)
{
  ros::init(argc, argv, "adder_server");
  ros::NodeHandle n;

  ros::ServiceServer service =
n.advertiseService("chapter2_tutorials/adder", add);
  ROS_INFO("adder_server has started");
  ros::spin();

  return 0;
}
```

Let's discuss the code. These lines include the necessary headers and the `srv` file which was created previously:

```
#include "ros/ros.h"
#include "chapter2_tutorials/chapter2_srv.h"
```

The following function will add two variables and send the result to the client node:

```
bool add(chapter2_tutorials::chapter2_srv::Request  &req,
         chapter2_tutorials::chapter2_srv::Response &res)
{
  res.sum = req.A + req.B;
  ROS_INFO("Request: A=%d, B=%d", (int)req.A, (int)req.B);
```

```
  ROS_INFO("Response: [%d]", (int)res.sum);
  return true;
}
```

Here, the service is created and advertised over the ROS computational network:

```
ros::ServiceServer service = n.advertiseService("chapter2_tutorials/adder",
add);
```

In the second file, `example_3b.cpp`, we will add this code:

```
#include "ros/ros.h"
#include "chapter2_tutorials/chapter2_srv.h"
#include <cstdlib>

int main(int argc, char **argv)
{
  ros::init(argc, argv, "adder_client");
  if (argc != 3)
  {
    ROS_INFO("Usage: adder_client A B ");
    return 1;
  }

  ros::NodeHandle n;
  ros::ServiceClient client =
n.serviceClient<chapter2_tutorials::chapter2_srv>("chapter2_tutorials/adder
");
  chapter2_tutorials::chapter2_srv srv;
  srv.request.A = atoll(argv[1]);
  srv.request.B = atoll(argv[2]);
  if (client.call(srv))
  {
    ROS_INFO("Sum: %ld", (long int)srv.response.sum);
  }
  else
  {
    ROS_ERROR("Failed to call service adder_server");
    return 1;
  }

  return 0;
}
```

We will create a client for the service with the name chapter2_tutorials/adder:

```
ros::ServiceClient client =
n.serviceClient<chapter2_tutorials::chapter2_srv>("chapter2_tutorials/adder
");
```

In the following code, we will create an instance of our srv request type and fill all the values to be sent, which has two fields:

```
chapter2_tutorials::chapter2_srv srv;
srv.request.A = atoll(argv[1]);
srv.request.B = atoll(argv[2]);
```

In the next line, the service is called and the data is sent. If the call succeeds, call() will return true, otherwise, call() will return false:

```
if (client.call(srv))
```

To build the service and client nodes created just now, we will have to add the following lines in CMakeList.txt:

```
add_executable(example3a src/example_3a.cpp)
add_executable(example3b src/example_3b.cpp)

add_dependencies(example3a chapter2_tutorials_generate_messages_cpp)
add_dependencies(example3b chapter2_tutorials_generate_messages_cpp)
target_link_libraries(example3a ${catkin_LIBRARIES})
target_link_libraries(example3b ${catkin_LIBRARIES})
```

The catkin_make tool is used to build the package, which will compile all the nodes:

```
$ cd ~/catkin_ws
$ catkin_make
```

To work with these nodes, we will have to execute the following commands in two separate shells:

```
$ rosrun chapter2_tutorials example3a
$ rosrun chapter2_tutorials example3b 2 3
```

The output will be as follows:

```
kbipin@ubuntu:~$ rosrun chapter2_tutorials example3a
[ INFO] [1500802889.035937665]: adder_server has started
[ INFO] [1500802898.176382487]: Request: A=2, B=3
[ INFO] [1500802898.176455696]: Response: [5]

bipin@ubuntu:~$ rosrun chapter2_tutorials example3b 2 3
[NFO] [1500802898.177676886]: Sum: 5
```

Server and client

See also

We have looked into the basic concepts of working with ROS and learned about publisher and subscriber, client and server, and parameter servers. In this section, we will learn about using advanced tools with ROS.

Understanding the ROS launch file

In previous sections, we created nodes and have been executing them in different shells. Imagine working with 20 nodes and the nightmare of executing each one in a shell! However, with the `launch` file, we can do it in the same shell by launching a configuration file with the extension `.launch`. The launch file is a useful feature in ROS for launching more than one node.

To learn about the ROS launch file, we will create a new folder in the `chapter2_tutorials` package, as follows:

```
$ roscd chapter2_tutorials/
$ mkdir launch
$ cd launch
$ vim chapter2.launch
```

Add the following code inside the `chapter2.launch` file:

```
<?xml version="1.0"?>
<launch>
    <node name ="example1a" pkg="chapter2_tutorials" type="example1a"
output="screen"/>
    <node name ="example1b" pkg="chapter2_tutorials" type="example1b"
output="screen"/>
</launch>
```

This file is simple, although we would have to write a very complex file to, for example, control a complete robot, such as **PR2**. These could be real robots or simulated in ROS.

The ROS launch is an XML file with the `.launch` extension and has a launch tag. Inside this tag, we will find the `node` tag, which is used to launch a node from a package, for example, the `example1a` node from the `chapter2_tutorials` package.

The `chapter2.launch` file will execute two nodes—the first two examples of this chapter :`example1a` and `example1b`.

To launch the file, use the following command:

```
$ roslaunch chapter2_tutorials chapter2.launch
```

We will see something similar to the following screenshot on our screens:

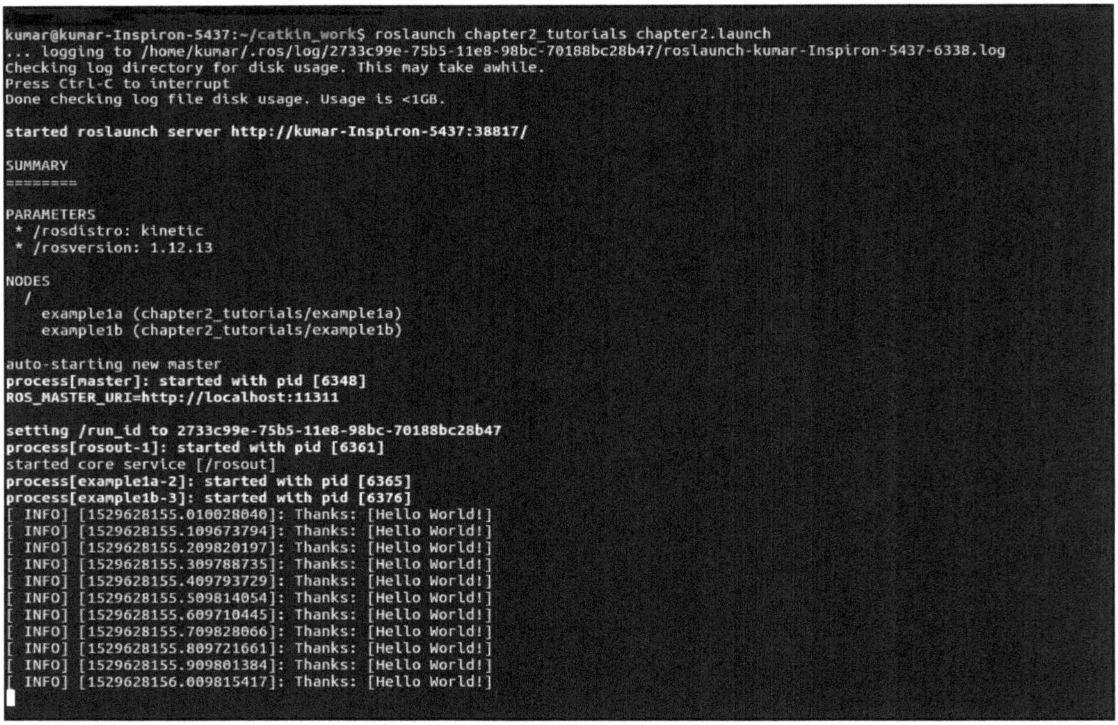

```
kumar@kumar-Inspiron-5437:~/catkin_work$ roslaunch chapter2_tutorials chapter2.launch
... logging to /home/kumar/.ros/log/2733c99e-75b5-11e8-98bc-70188bc28b47/roslaunch-kumar-Inspiron-5437-6338.log
Checking log directory for disk usage. This may take awhile.
Press Ctrl-C to interrupt
Done checking log file disk usage. Usage is <1GB.

started roslaunch server http://kumar-Inspiron-5437:38817/

SUMMARY
========

PARAMETERS
 * /rosdistro: kinetic
 * /rosversion: 1.12.13

NODES
  /
    example1a (chapter2_tutorials/example1a)
    example1b (chapter2_tutorials/example1b)

auto-starting new master
process[master]: started with pid [6348]
ROS_MASTER_URI=http://localhost:11311

setting /run_id to 2733c99e-75b5-11e8-98bc-70188bc28b47
process[rosout-1]: started with pid [6361]
started core service [/rosout]
process[example1a-2]: started with pid [6365]
process[example1b-3]: started with pid [6376]
[ INFO] [1529628155.010028040]: Thanks: [Hello World!]
[ INFO] [1529628155.109673794]: Thanks: [Hello World!]
[ INFO] [1529628155.209820197]: Thanks: [Hello World!]
[ INFO] [1529628155.309788735]: Thanks: [Hello World!]
[ INFO] [1529628155.409793729]: Thanks: [Hello World!]
[ INFO] [1529628155.589814054]: Thanks: [Hello World!]
[ INFO] [1529628155.609710445]: Thanks: [Hello World!]
[ INFO] [1529628155.709828066]: Thanks: [Hello World!]
[ INFO] [1529628155.809721661]: Thanks: [Hello World!]
[ INFO] [1529628155.909801384]: Thanks: [Hello World!]
[ INFO] [1529628156.009815417]: Thanks: [Hello World!]
```

ROS launch

The running nodes are listed in the screenshot. When you launch a launch file, it is not necessary to execute it before the `roscore` command; `roslaunch` does it for us.

3
ROS Architecture and Concepts – II

In this chapter, we will discuss the following recipes:

- Understanding the parameter server and dynamic parameters
- Understanding the ROS actionlib
- Understanding the ROS pluginlib
- Understanding the ROS nodelets
- Understanding the Gazebo framework and plugin
- Understanding the ROS transform frame (TF)
- Understanding the ROS Visualization tool (RViz) and its plugins

Introduction

In the previous chapter, we looked into the general information about the ROS architecture and how it works. We saw certain concepts, tools, and samples of how to interact with nodes, topics, and services. However, in the beginning, all of these concepts might look complicated, but, in the upcoming chapters, we will start to understand their applications. It is advisable to practice these terms and refer to tutorials before reading the upcoming chapters.

In this chapter, we will learn the advanced concepts of the ROS architecture, which includes the parameter server and dynamic parameter, `actionlib`, and `nodelets`, along with TF broadcaster and listener. In addition, we will also look into the ROS simulation framework, *Gazebo*, and how to develop a plugin for it. Finally, we will look at the **ROS visualization tool** (**RViz**) and its plugins. We will discuss the functionalities and applications of each concept and will look at an example to demonstrate its workings.

Understanding the parameter server and dynamic parameters

In the previous chapter, we learned that the parameter server is a part of the ROS Master and allows the ROS system to keep the data or configuration information that is to be stored in a central place. All nodes can access and modify these values.

We have experience with the `rosparam` tool so we can work with the parameter server. The parameters can be changed dynamically during the execution of the node that uses these parameters, using the `dyamic_reconfigure` package. In the next section, we will learn about the dynamic parameters utility in ROS in detail.

Getting ready

Generally, we program a node where we initialize the variables with data values that can only be modified within the node. If it is required to modify these values dynamically from outside the running node, we can use the Parameter Server, services, or topics.

For example, if we are working with a node that uses the PID controller to control the optimum speed of the motor, it is often required to tune the PID parameters, $k1$, $k2$, and $k3$. However, ROS has the `Dynamic Reconfigure` utility to perform these functions more efficiently.

In the following section, we will learn how to enable this feature in a basic example node, which requires adding the necessary lines in the `CMakeLists.txt` and `package.xml` files.

How to do it...

1. To use the `Dynamic Reconfigure` utility, we will write a configuration file and save it in the `cfg` folder in the intended package.
2. Create the `cfg` folder and a `parameter_server_tutorials.cfg` file there in the `parameter_server_tutorials` package, as follows:

```
$ roscd parameter_server_tutorials
$ mkdir cfg
$ vim parameter_server_tutorials
```

3. We will add the following code in the `parameter_server_tutorials.cfg` file:

```python
#!/usr/bin/env python
PACKAGE = "parameter_server_tutorials"

from dynamic_reconfigure.parameter_generator_catkin import *

gen = ParameterGenerator()

gen.add("BOOL_PARAM", bool_t, 0,"A Boolean  parameter", True)
gen.add("INT_PARAM", int_t, 0, "An Integer Parameter", 1, 0,
100)
gen.add("DOUBLE_PARAM", double_t, 0, "A Double  Parameter",
0.01, 0,    1)
gen.add("STR_PARAM", str_t, 0, "A String  parameter", "Dynamic
Reconfigure")

size_enum = gen.enum([ gen.const("Low", int_t, 0, "Low : 0"),
                       gen.const("Medium", int_t, 1,"Medium :
1"),
                       gen.const("High", int_t, 2, "Hight
:2")],
                       "Selection List")
gen.add("SIZE", int_t, 0, "Selection List", 1, 0, 3,
edit_method=size_enum)

exit(gen.generate(PACKAGE, "parameter_server_tutorials",
"parameter_server_"))
```

4. To initialize the parameter generator and add parameters in the following few lines, add the following:

```python
gen = ParameterGenerator()
gen.add("BOOL_PARAM", bool_t, 0,"A Boolean  parameter",  True)
gen.add("INT_PARAM", int_t, 0, "An Integer Parameter", 1, 0,
100)
gen.add("DOUBLE_PARAM", double_t, 0, "A Double Parameter",
0.01, 0,    1)
gen.add("STR_PARAM", str_t, 0, "A String parameter", "Dynamic
Reconfigure")
size_enum = gen.enum([gen.const("Low", int_t, 0, "Low : 0"),
                       gen.const("Medium", int_t, 1, "Medium :
1"),
                       gen.const("High", int_t,  2, "Hight
:2")],"Selection List")
gen.add("SIZE", int_t, 0, "Selection List", 1, 0, 3,
edit_method=size_enum)
```

5. These lines add different parameter types and set the default values, description, range, and so on. Moreover, the parameter has the following arguments:

```
gen.add(name, type, level, description, default, min, max)
```

The arguments are defined as follows:

- `name`: The name of the parameter
- `type`: The type of value stored
- `level`: The bitmask that is passed to the callback
- `description`: The description of the parameter
- `default`: The default value when the node starts
- `min`: The minimum value for the parameter
- `max`: The maximum value for the parameter

Finally, the last line of the previous code generates the necessary files and exits the program. Here, we can see that the `.cfg` file is written in Python. However, this book is for C++, but sometimes we will use Python whenever it is appropriate and it is required to explain a certain concept:

```
exit(gen.generate(PACKAGE, "parameter_server_tutorials",
"parameter_server_"))
```

It is necessary to change the permissions for the `parameter_server_tutorials.cfg` file because the file will be executed by ROS as Python scripts:

```
$ chmod a+x cfg/ parameter_server_tutorials.cfg
```

6. For invoking the compilation, the following lines should be added in `CMakeList.txt`:

```
find_package(catkin REQUIRED COMPONENTS
  roscpp
  std_msgs
  message_generation
  dynamic_reconfigure
)

generate_dynamic_reconfigure_options(
  cfg/parameter_server_tutorials.cfg
)
add_dependencies(parameter_server_tutorials
parameter_server_tutorials_gencfg)
```

7. Create an example node with the Dynamic Reconfigure support:

```
$ roscd parameter_server_tutorials
$ vim src/ parameter_server_tutorials.cpp
```

We will add the following code snippet in the node file:

```
#include <ros/ros.h>
#include <dynamic_reconfigure/server.h>
#include <parameter_server_tutorials/parameter_server_Config.h>

void
callback(parameter_server_tutorials::parameter_server_Config
&config, uint32_t level) {

  ROS_INFO("Reconfigure Request: %s %d %f %s %d",
          config.BOOL_PARAM?"True":"False",
          config.INT_PARAM,
          config.DOUBLE_PARAM,
          config.STR_PARAM.c_str(),
          config.SIZE);

}

int main(int argc, char **argv) {
  ros::init(argc, argv, "parameter_server_tutorials");

dynamic_reconfigure::Server<parameter_server_tutorials::paramet
er_server_Config> server;
dynamic_reconfigure::Server<parameter_server_tutorials::paramet
er_server_Config>::CallbackType f;

  f = boost::bind(&callback, _1, _2);
  server.setCallback(f);

  ROS_INFO("Spinning");
  ros::spin();
  return 0;
}
```

As usual, these lines include the headers for ROS, the parameter server, and the config file that we created earlier:

```
#include <ros/ros.h>
#include <dynamic_reconfigure/server.h>
#include <parameter_server_tutorials/parameter_server_Config.h>
```

The following lines of code show the `callback` function, which will print the values for the parameters that are modified by the user. Notice that the name of the parameter must be the same as the one that is configured in the `parameter_server_tutorials.cfg` file:

```
void
callback(parameter_server_tutorials::parameter_server_Config
&config, uint32_t level) {

  ROS_INFO("Reconfigure Request: %s %d %f %s %d",
           config.BOOL_PARAM?"True":"False",
           config.INT_PARAM,
           config.DOUBLE_PARAM,
           config.STR_PARAM.c_str(),
           config.SIZE);

}
```

Moreover, in the `main` function, the server is initialized with the `parameter_server_Config configuration` file. The next `callback` function is set to the server which will be called, which is when the server gets a reconfiguration request:

```
dynamic_reconfigure::Server<parameter_server_tutorials::paramet
er_server_Config> server;
dynamic_reconfigure::Server<parameter_server_tutorials::paramet
er_server_Config>::CallbackType f;

  f = boost::bind(&callback, _1, _2);
  server.setCallback(f);
```

8. Finally, we have to add the following lines to the `CMakeLists.txt` file for the ROS build system to invoke the compilation:

```
add_executable(parameter_server_tutorials
src/parameter_server_tutorials.cpp)

add_dependencies(parameter_server_tutorials
parameter_server_tutorials_gencfg)

target_link_libraries(parameter_server_tutorials
${catkin_LIBRARIES}
```

Great! We are done with the development part. Now, we have to compile and run the node and the **Dynamic Reconfigure** GUI as follows:

```
$ roscore
$ rosrun parameter_server_tutorials parameter_server_tutorials
$ rosrun rqt_reconfigure rqt_reconfigure
```

9. To dynamically modify the parameters of the node, as shown in the following screenshot:

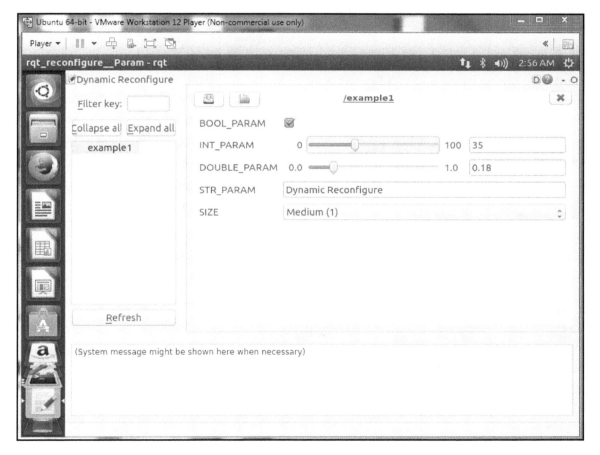

The Dynamic Reconfigure utility GUI

Each time a user modifies a parameter with the slider, the checkbox, and so on, we will observe the changes made in the shell where the node is running:

```
^Ckbipin@ubuntu:~$ rosrun chapter3_tutorials example1
[ INFO] [1502444374.797267810]: Reconfigure Request: True 1 0.010000 Dynamic Reconfigure 1
[ INFO] [1502444374.812956575]: Spinning
[ INFO] [1502444437.136884221]: Reconfigure Request: True 3 0.010000 Dynamic Reconfigure 1
[ INFO] [1502444438.914553379]: Reconfigure Request: False 3 0.010000 Dynamic Reconfigure 1
[ INFO] [1502444444.149008958]: Reconfigure Request: False 3 0.010000 Dynamic Reconfigure 2
[ INFO] [1502444445.592562578]: Reconfigure Request: False 3 0.020000 Dynamic Reconfigure 2
[ INFO] [1502444448.205849872]: Reconfigure Request: False 3 0.070000 Dynamic Reconfigure 2
```

The Dynamic Reconfigure utility callback

Thanks to the `Dynamic Reconfigure` utility of ROS, the node which interfaces with the hardware can be tuned and validated more efficiently and faster. We will learn more about this in the upcoming chapters.

Understanding the ROS actionlib

In the ROS computational network, there could be several cases where someone needs to send a request to a node to perform a task, and also receive a reply to that request. This can be possible via ROS services.

Nevertheless, in some cases, if the service takes a long time to execute or a requested service is not valid now, the client might want the ability to preempt the request during execution and get periodic feedback via a request progress status. The ROS tools provide the `actionlib` package to create servers that execute long-running goals that could be preempted or resubmitted.

Getting ready

The `actionlib` package consists of `ActionClient` and `ActionServer`, which communicate via a "ROS Action Protocol" built on top of ROS messages via function calls and callbacks.

The following diagram shows the interaction between the client and server applications:

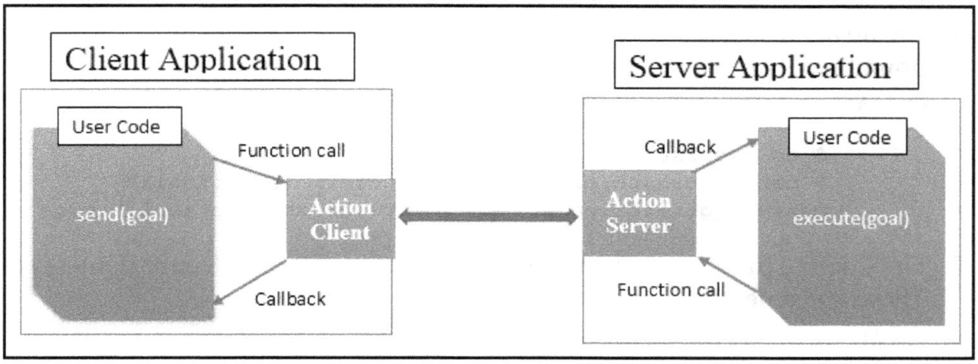

Client-server interaction

We have to define a few messages that are action-specific, through which the client and server can communicate.

These are defined as goal, feedback, and result messages:

- **Goal**: We will introduce the notion of a goal as tasks that are accomplished using actions which can be sent to `ActionServer` by `ActionClient`. For example, in the case of a mobile robot, the goal would be to choose the next pose (*x, y, z, phi, chi, theta*), which has information about where the mobile robot should move to in the world.
- **Feedback**: `ActionServer` provides a feedback message to `ActionClient`, which defines a way to tell the incremental progress of a goal periodically. In the case of the mobile robot, this might be the current pose of the robot, along with other information.
- **Result**: Upon completion of this goal, `ActionServer` sends the result message to `ActionClient`, which is quite different from a feedback message, since it is sent exactly once and is extremely useful in some cases. However, in the case of the mobile robot, the result isn't very important, but it might contain the final pose of the robot.

In the following section, we will learn how to create an action server and client, where the `ActionServer` generates a Fibonacci sequence, the goal is the order of the sequence, the feedback is the sequence as it is computed, and the result is the final sequence. This also includes `ActionClient`, which sends a goal to the action server.

How to do it...

1. Create an `actionlib_tutorials` package with the following dependencies or download it from the GitHub repository (`https://github.com/kbipin/Robot-Operating-System-Cookbook`):

   ```
   $ cd <workspace>/src
   $ catkin_create_pkg actionlib_tutorials actionlib
   message_generation roscpp rospy std_msgs actionlib_msgs
   ```

2. First, define the action which consists of goal, result, and feedback messages. These are generated automatically from the `.action` file.

3. Create an `actionlib_tutorials/action/Fibonacci.action` file with the following contents:

   ```
   #goal definition
   int32 order
   ---
   #result definition
   int32[] sequence
   ---
   #feedback
   int32[] sequence
   ```

4. To automatically generate the message files during the creation process, the following elements need to be added to `CMakeLists.txt`:

   ```
   find_package(catkin REQUIRED COMPONENTS
     actionlib
     actionlib_msgs
     message_generation
     roscpp
     rospy
     std_msgs
   )



   add_action_files(
     DIRECTORY action
     FILES Fibonacci.action
   )

   generate_messages(
   DEPENDENCIES actionlib_msgs std_msgs
   ```

```
)

catkin_package(
  INCLUDE_DIRS include
  LIBRARIES actionlib_tutorials
  CATKIN_DEPENDS actionlib actionlib_msgs message_generation
roscpp rospy std_msgs
  DEPENDS system_lib
)
add_executable(fibonacci_server src/fibonacci_server.cpp)

target_link_libraries(
  fibonacci_server
  ${catkin_LIBRARIES}
)

add_dependencies(
  fibonacci_server
  ${actionlib_tutorials_EXPORTED_TARGETS}
)
add_executable(fibonacci_client src/fibonacci_client.cpp)

target_link_libraries(
  fibonacci_client
  ${catkin_LIBRARIES}
)

add_dependencies(
  fibonacci_client
  ${actionlib_tutorials_EXPORTED_TARGETS}
)
```

The `catkin_make` command will automatically generate the required messages and header files, as shown in the following screenshot:

```
kumar@kumar-Inspiron-5437:~/catkin_ws/devel/share/actionlib_tutorials/msg$ ls
FibonacciActionFeedback.msg  FibonacciAction.msg        FibonacciFeedback.msg  FibonacciResult.msg
FibonacciActionGoal.msg      FibonacciActionResult.msg  FibonacciGoal.msg
kumar@kumar-Inspiron-5437:~/catkin_ws/devel/share/actionlib_tutorials/msg$
```

Action messages and file generation

The following code snippet of the action server in the `actionlib_tutorials/src/fibonacci_server.cpp` file is self-explanatory:

```
#include <ros/ros.h>
#include <actionlib/server/simple_action_server.h>
#include <actionlib_tutorials/FibonacciAction.h>
```

```
class FibonacciAction
{
protected:

  ros::NodeHandle nh_;
  /* NodeHandle instance must be created before this line.
Otherwise strange error occurs.*/
actionlib::SimpleActionServer<actionlib_tutorials::FibonacciAct
ion> as_;
  std::string action_name_;
  /* create messages that are used to published feedback/result
*/
  actionlib_tutorials::FibonacciFeedback feedback_;
  actionlib_tutorials::FibonacciResult result_;

public:

  FibonacciAction(std::string name) :
    as_(nh_, name, boost::bind(&FibonacciAction::executeCB,
this, _1), false),
    action_name_(name)
  {
    as_.start();
  }

  ~FibonacciAction(void)
  {
  }

  void executeCB(const
actionlib_tutorials::FibonacciGoalConstPtr &goal)
  {
    ros::Rate r(1);
    bool success = true;

    /* the seeds for the fibonacci sequence */
    feedback_.sequence.clear();
    feedback_.sequence.push_back(0);
    feedback_.sequence.push_back(1);

    ROS_INFO("%s: Executing, creating fibonacci sequence of
order %i with seeds %i, %i", action_name_.c_str(), goal->order,
feedback_.sequence[0], feedback_.sequence[1]);

    /* start executing the action */
    for(int i=1; i<=goal->order; i++)
    {
     /* check that preempt has not been requested by the client
```

```
*/
        if (as_.isPreemptRequested() || !ros::ok())
        {
          ROS_INFO("%s: Preempted", action_name_.c_str());
          /* set the action state to preempted */
          as_.setPreempted();
          success = false;
          break;
        }
        feedback_.sequence.push_back(feedback_.sequence[i] +
feedback_.sequence[i-1]);
      /* publish the feedback */
        as_.publishFeedback(feedback_);
        /* this sleep is not necessary, however, the sequence is
computed at 1 Hz for demonstration purposes */
        r.sleep();
      }
      if(success)
      {
        result_.sequence = feedback_.sequence;
        ROS_INFO("%s: Succeeded", action_name_.c_str());
       /* set the action state to succeeded */
        as_.setSucceeded(result_);
      }
    }
};

int main(int argc, char** argv)
{
  ros::init(argc, argv, "fibonacci server");

  FibonacciAction fibonacci("fibonacci");
  ros::spin();

  return 0;
}
```

Now, let's discuss the important parts of this code in detail:

```
FibonacciAction(std::string name) :
    as_(nh_, name, boost::bind(&FibonacciAction::executeCB,
this, _1), false),
    action_name_(name)
  {
    as_.start();
  }
```

In the constructor of the `FibonacciAction` class, an action server is initialized and started, which takes a node handle, the name of the action, and optionally, a callback as arguments. In this example, the action server is created with `executeCB` being used as a `callback` function:

```
/* start executing the action */
    for(int i=1; i<=goal->order; i++)
    {
     /* check that preempt has not been requested by the client
*/
        if (as_.isPreemptRequested() || !ros::ok())
        {
          ROS_INFO("%s: Preempted", action_name_.c_str());
          /* set the action state to preempted */
          as_.setPreempted();
          success = false;
          break;
        }
        feedback_.sequence.push_back(feedback_.sequence[i] +
feedback_.sequence[i-1]);
        /* publish the feedback */
        as_.publishFeedback(feedback_);
        /* this sleep is not necessary, however, the sequence is
computed at 1 Hz for demonstration purposes */
        r.sleep();
    }
```

One of the most important features of the action server is the ability to allow an action client to preempt the current goal execution. When a client requests the preemption of the current goal, the action server will cancel the goal execution, perform essential clean-up, and call the `setPreempted()` function, which informs the ROS framework that the action has been preempted by a client request. The rate at which the action server checks for preemption requests and provides the feedback information is implementation-dependent:

```
if(success)
    {
        result_.sequence = feedback_.sequence;
        ROS_INFO("%s: Succeeded", action_name_.c_str());
       /* set the action state to succeeded */
        as_.setSucceeded(result_);
    }
```

Once the action server has finished executing its current goal, it informs the action client about completion by calling the `setSucceeded()` function. From now on, the action server is running and waiting to receive its next set of goals.

The following code snippet of the action client's
`actionlib_tutorials/src/fibonacci_client.cpp` file is self-explanatory:

```cpp
#include <ros/ros.h>
#include <actionlib/client/simple_action_client.h>
#include <actionlib/client/terminal_state.h>
#include <actionlib_tutorials/FibonacciAction.h>

int main (int argc, char **argv)
{
  ros::init(argc, argv, "fibonacci client");

  /* create the action client
     "true" causes the client to spin its own thread */
actionlib::SimpleActionClient<actionlib_tutorials::FibonacciAct
ion> ac("fibonacci", true);

  ROS_INFO("Waiting for action server to start.");
  /* will be  waiting for infinite time */
  ac.waitForServer();

  ROS_INFO("Action server started, sending goal.");

  actionlib_tutorials::FibonacciGoal goal;
  goal.order = 20;
  ac.sendGoal(goal);

  /* waiting for the action to return */
  bool finished_before_timeout =
ac.waitForResult(ros::Duration(30.0));

  if (finished_before_timeout)
  {
    actionlib::SimpleClientGoalState state = ac.getState();
    ROS_INFO("Action finished: %s",state.toString().c_str());
  }
  else
    ROS_INFO("Action does not finish before the time out.");

  return 0;
}
```

Now, let's discuss the key elements of the code in detail:

```
// create the action client
// true causes the client to spin its own thread

actionlib::SimpleActionClient<actionlib_tutorials::FibonacciAct
ion> ac("fibonacci", true);
ROS_INFO("Waiting for action server to start.");
// wait for the action server to start

  ac.waitForServer(); //will wait for infinite time
```

Here, the action client is constructed with the server name and the auto spin option set to `true` which takes two arguments, the action server name and a boolean option to automatically spin a thread (`actionlib` is used to do the 'thread magic' behind the scenes). It also specifies message types to communicate with the action server.

In the following line, the action client will wait for the action server to start before continuing:

```
// send a goal to the action
  actionlib_tutorials::FibonacciGoal goal;
  goal.order = 20;
  ac.sendGoal(goal);
  //wait for the action to return
  bool finished_before_timeout =
ac.waitForResult(ros::Duration(30.0));
  if (finished_before_timeout)
  {
    actionlib::SimpleClientGoalState state = ac.getState();
    ROS_INFO("Action finished: %s",state.toString().c_str());
  }
  else
    ROS_INFO("Action did not finish before the time out.");
```

Here, the action client creates the goal message and sends it to the action server. It now waits for the goal to finish before continuing. The timeout on the wait is set to 30 seconds, which means that the function will return with false if the goal has not finished and the user is notified that the goal did not finish in the allotted time. If the goal finishes before a timeout, the goal status is reported and normal execution continues.

5. To run the following commands in the top level directory of our workspace to compile the package, , execute the following command:

   ```
   $ catkin_make
   ```

6. To set up the system environment, let's execute the following command:

   ```
   $ source devel/setup.bash
   ```

   After compiling the executable successfully, we will start the action client and server in a separate terminal, assuming that roscore is running:

   ```
   $ rosrun actionlib_tutorials fibonacci_server
   ```

   The output is shown in the following screenshot:

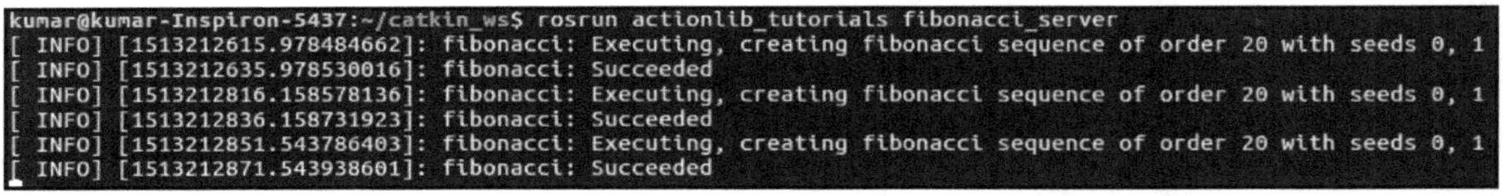

```
kumar@kumar-Inspiron-5437:~/catkin_ws$ rosrun actionlib_tutorials fibonacci_server
[ INFO] [1513212615.978484662]: fibonacci: Executing, creating fibonacci sequence of order 20 with seeds 0, 1
[ INFO] [1513212635.978530016]: fibonacci: Succeeded
[ INFO] [1513212816.158578136]: fibonacci: Executing, creating fibonacci sequence of order 20 with seeds 0, 1
[ INFO] [1513212836.158731923]: fibonacci: Succeeded
[ INFO] [1513212851.543786403]: fibonacci: Executing, creating fibonacci sequence of order 20 with seeds 0, 1
[ INFO] [1513212871.543938601]: fibonacci: Succeeded
```

Action server outputs

Next, we would start the action client using following command:
```
$ rosrun actionlib_tutorials fibonacci_client
```

The output is shown in the following screenshot:

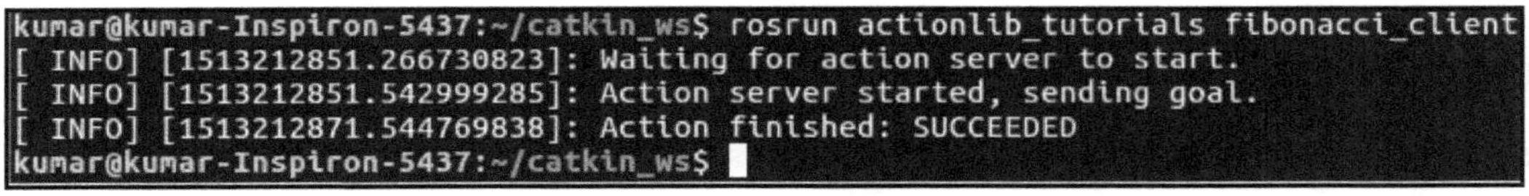

```
kumar@kumar-Inspiron-5437:~/catkin_ws$ rosrun actionlib_tutorials fibonacci_client
[ INFO] [1513212851.266730823]: Waiting for action server to start.
[ INFO] [1513212851.542999285]: Action server started, sending goal.
[ INFO] [1513212871.544769838]: Action finished: SUCCEEDED
kumar@kumar-Inspiron-5437:~/catkin_ws$
```

Action client outputs

7. To visualize the action server and client execution graphically, execute the following command:

   ```
   $ rqt_graph
   ```

The output is shown in the following screenshot:

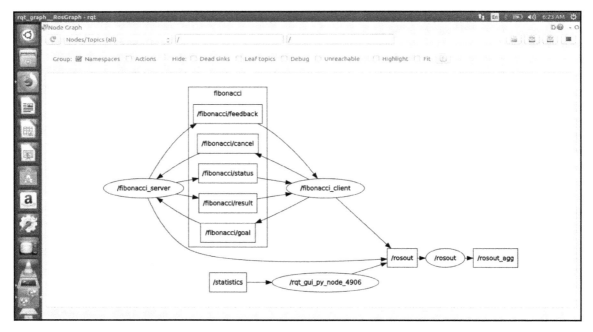

Action client computation graph

The preceding screenshot shows that the action server is publishing the feedback, status, and result channels and is subscribed to the goal and cancel channels. Similarly, the action client is subscribing to the feedback, status, and result channels and is publishing to the goal and cancel channels. Both the server and client are up and running as expected.

To see feedback from the action server while it's acting on its goal, run the following command:

```
$ rostopic echo /fibonacci/feedback
```

The output is shown in the following screenshot:

```
kumar@kumar-Inspiron-5437:~/catkin_ws$ rostopic echo /fibonacci/feedback
header:
  seq: 44
  stamp:
    secs: 1513212855
    nsecs: 543885567
  frame_id: ''
status:
  goal_id:
    stamp:
      secs: 1513212851
      nsecs: 543089279
    id: /fibonacci_client-1-1513212851.543089279
  status: 1
  text: This goal has been accepted by the simple action server
feedback:
  sequence: [0, 1, 1, 2, 3, 5, 8]
---
```

Feedback from the action server

Similarly, in a new terminal, we can observe the feedback channel to see the result from the action server after the goal has been completed:

```
kumar@kumar-Inspiron-5437:~/catkin_ws$ rostopic echo /fibonacci/result
header:
  seq: 3
  stamp:
    secs: 1513213621
    nsecs: 978107629
  frame_id: ''
status:
  goal_id:
    stamp:
      secs: 1513213601
      nsecs: 977271538
    id: /fibonacci_client-1-1513213601.977271538
  status: 3
  text: ''
result:
  sequence: [0, 1, 1, 2, 3, 5, 8, 13, 21, 34, 55, 89, 144, 233, 377, 610, 987, 1597, 2584, 4181, 6765, 10946]
---
```

Results from the action server

Understanding the ROS pluginlib

Software plugins are a well-known approach in software applications, and are used as add-ons for the existing application and to provide some new features. Moreover, the plugins enable an infrastructure for the software architecture development to accept new features whenever required. In other words, using this method, which is highly modular, we can enhance the capabilities of the software to any level in the future and deliver this to the user in the form of updates in runtime without modifying the main application.

The software plugins are shared objects (.so) or dynamic link libraries that do not have any dependencies with the main software and are built without having to link to the main application code.

Getting ready

A complex robotic application also uses the ROS `pluginlib` framework to extend the capabilities of the robot. Here, we will develop a simple `RegularPolygon` application using `pluginlib` and understand all the procedures involved in creating and linking a plugin to the main application.

First of all, we will create `pluginlib_tutorials` inside our workspace:

```
$ catkin_create_pkg pluginlib_tutorials roscpp pluginlib
```

How to do it...

First, we will implement an abstract base class of polygon which can be extended to any type of polygons such as triangles, or even squares. In the future, we could add this to any complex polygon by writing an individual plugin for each one.

Creating plugins

We will create a base class called `RegularPolygon` from which all of our plugins, including `Triangle` and `Square`, would inherit.

The source code of packages can be downloaded from GitHub (https://github.com/
kbipin/Robot-Operating-System-Cookbook):

```
#ifndef PLUGINLIB_TUTORIALS__POLYGON_BASE_H_
#define PLUGINLIB_TUTORIALS__POLYGON_BASE_H_

namespace polygon_base
{
  class RegularPolygon
  {
    public:
      virtual void initialize(double side_length) = 0;
      virtual double area() = 0;
      virtual ~RegularPolygon(){}

    protected:
      RegularPolygon(){}
  };
};
#endif
```

We will create two RegularPolygon plugins; the first will be Triangle and the second
will be Square:
pluginlib_tutorials/include/pluginlib_tutorials/polygon_plugins.h.

```
#ifndef PLUGINLIB_TUTORIALS__POLYGON_PLUGINS_H_
#define PLUGINLIB_TUTORIALS__POLYGON_PLUGINS_H_
#include <pluginlib_tutorials/polygon_base.h>
#include <cmath>
namespace polygon_plugins
{
  class Triangle : public polygon_base::RegularPolygon
  {
    public:
      Triangle(){}
      void initialize(double side_length)
      {
        side_length_ = side_length;
      }
      double area()
      {
        return 0.5 * side_length_ * getHeight();
      }
      double getHeight()
      {
        return sqrt((side_length_ * side_length_) - ((side_length_ / 2)
  * (side_length_ / 2)));
```

```
      }
    private:
      double side_length_;
  };
  class Square : public polygon_base::RegularPolygon
  {
    public:
      Square(){}
      void initialize(double side_length)
      {
        side_length_ = side_length;
      }
      double area()
      {
        return side_length_ * side_length_;
      }
    private:
      double side_length_;
  };
};
#endif
```

The preceding code should be self-explanatory.

Compiling the plugin library

We have to add the following line in the CMakeLists.txt file to compile the library:

```
include_directories(include)
add_library(polygon_plugins src/polygon_plugins.cpp)
```

Plugins registration

In the preceding subsection, we created standard C++ classes. Now, we will discuss pluginlib-specific work as a declaration of the Triangle and Square classes as plugins. Let's look at the source code in the src/polygon_plugins.cpp file:

```
#include <pluginlib/class_list_macros.h>
#include <pluginlib_tutorials_/polygon_base.h>
#include <pluginlib_tutorials_/polygon_plugins.h>

PLUGINLIB_EXPORT_CLASS(polygon_plugins::Triangle,
polygon_base::RegularPolygon)
PLUGINLIB_EXPORT_CLASS(polygon_plugins::Square,
polygon_base::RegularPolygon)
```

Here, we have registered the `Triangle` and `Square` classes as plugins using the `PLUGINLIB_EXPORT_CLASS` macro.

Making the plugins available to the ROS toolchain

After compilation, the plugins will be available as a share library. The ROS toolchain plugin loader requires a bit of information for linking and resolving the reference within the library. This information will be provided by `polygon_plugins.xml` in packages along with a special export line in the package manifest.

The Plugin XML File

Here's the XML:

```xml
<library path="lib/libpolygon_plugins">
  <class type="polygon_plugins::Triangle"
base_class_type="polygon_base::RegularPolygon">
    <description>This is a triangle plugin.</description>
  </class>
  <class type="polygon_plugins::Square"
base_class_type="polygon_base::RegularPolygon">
    <description>This is a square plugin.</description>
  </class>
</library>
```

This is where the `library` tag provides the relative path of the library that contains the plugins. Similarly, the `class` tag declares a plugin that will be exported from the plugins library.

Exporting plugins

As mentioned in the previous section, to export the plugins, we have to add the following lines to the `polygon_plugins.xml` package manifest file:

```xml
<export>
  <pluginlib_tutorials_ plugin="${prefix}/polygon_plugins.xml" />
</export>
```

To check that everything we have done is working, we will first need to build the workspace and source the resulting setup file, then try running the following `rospack` command:

```
$ catkin_make
$ rospack plugins --attrib=plugin pluginlib_tutorials
```

This should output the full path to the `polygon_plugins.xml` file if the ROS toolchain is set up properly to work with our plugin.

Using a plugin

In the previous section, we successfully created and exported a few of the `RegularPolygon` plugins. To discuss how to use them, we will create `src/polygon_loader.cpp` with the following contents:

```cpp
#include <pluginlib/class_loader.h>
#include <pluginlib_tutorials/polygon_base.h>

int main(int argc, char** argv)
{
  pluginlib::ClassLoader<polygon_base::RegularPolygon>
poly_loader("pluginlib_tutorials", "polygon_base::RegularPolygon");

  try
  {
    boost::shared_ptr<polygon_base::RegularPolygon> triangle =
poly_loader.createInstance("polygon_plugins::Triangle");
    triangle->initialize(10.0);

    boost::shared_ptr<polygon_base::RegularPolygon> square =
poly_loader.createInstance("polygon_plugins::Square");
    square->initialize(10.0);

    ROS_INFO("Triangle area: %.2f", triangle->area());
    ROS_INFO("Square area: %.2f", square->area());
  }
  catch(pluginlib::PluginlibException& ex)
  {
    ROS_ERROR("The plugin failed to load for some reason. Error: %s",
ex.what());
  }

  return 0;
}
```

The preceding source code loads the instance of `Triangle` and `Square` plugins belonging to `RegularPolygon` plugins and calculates the area. The `catch` block of code displays the reasons for failure if the `try` block fails.

Running the code

To build the source code discussed previously, we will add the following line of code to our `CMakeLists.txt` file and execute `catkin_make` from the top level directory:

```
$ rosrun pluginlib_tutorials polygon_loader
```

For the time being, we will not be able to visualize the usefulness of plugin development, however, in the upcoming section, it will become more visible.

Understanding the ROS nodelets

ROS provides nodelets as an execution unit; these are a type of ROS node that allow you to run multiple nodes in an umbrella of a single process. They correspond to threads and processes in the Linux system, and they're called **threaded nodes**. These could communicate with each other efficiently without overloading the network transport layer between two nodes. Moreover, these threaded nodes can also communicate with external nodes.

In the ROS framework, nodelets are used when the volume of data transferred among nodes is very high, for example, when transferring data from laser sensors or when a camera points to an object like a cloud, respectively.

Getting ready

In the previous section, we experienced the dynamic loading of a class with `pluginlib`. Similarly, in the case of nodelets, we will dynamically load each class as a plugin, which must have a separate namespace. Every loaded class can work as separate nodes, called nodelets, which execute on a single process.

How to do it...

In this section, we will develop a basic nodelet that will subscribe a topic (std_msgs/String) called /ros_in and publish the same message on the topic (std_msgs/String) called /ros_out. From there, we will discuss the workings of this nodelet.

Creating a nodelet

First of all, we will create a package called nodelet_hello_ros using the following command in our workspace:

```
$ catkin_create_pkg nodelet_hello_ros nodelet roscpp std_msgs
```

Here, the nodelet package provides APIs to build a ROS nodelet. Now, we will create a file called /src/hello_world.cpp, which contains the source code for nodelet implementation inside the package. Alternatively, we could use the existing package from chapter3_tutorials/nodelet_hello_ros on GitHub:

```
#include <pluginlib/class_list_macros.h>
#include <nodelet/nodelet.h>
#include <ros/ros.h>
#include <std_msgs/String.h>
#include <stdio.h>

namespace nodelet_hello_ros
{
class Hello : public nodelet::Nodelet
{
private:
   virtual void onInit()
   {
        ros::NodeHandle& private_nh = getPrivateNodeHandle();
        NODELET_DEBUG("Initialized Nodelet");
        pub = private_nh.advertise<std_msgs::String>("ros_out",5);
        sub = private_nh.subscribe("ros_in",5, &Hello::callback, this);
   }
   void callback(const std_msgs::StringConstPtr input)
   {

        std_msgs::String output;
        output.data = input->data;

        NODELET_DEBUG("msg data = %s",output.data.c_str());
```

```
            ROS_INFO("msg data = %s",output.data.c_str());
            pub.publish(output);
    }
    ros::Publisher pub;
    ros::Subscriber sub;
};
}
PLUGINLIB_DECLARE_CLASS(nodelet_hello_ros,Hello,nodelet_hello_ros::Hello,
nodelet::Nodelet);
```

Here, we will create a `nodelet` class called `Hello`, which is inherited from a standard `nodelet` base class. In the ROS framework, all `nodelet` classes should inherit from the `nodelet` base class and be dynamically loadable using `pluginlib`. Here, the `Hello` class will be dynamically loadable.

In the initialization function on the nodelet, we will create a node handle object, a publisher topic, `/ros_out`, and a subscriber on the topic, `/ros_in`. The subscriber is bound to a callback function called `callback()`. This is where we would print the messages from the `/ros_in` topic onto the console and publish to the `/ros_out` topic.

Finally, we will export the `Hello` class as a plugin for dynamic loading using the `PLUGINLIB_EXPORT_CLASS` macro.

Plugin description

Similar to `pluginlib`, we will create a plugin description file called `hello_ros.xml` inside the `nodelet_hello_ros` package with the following contents:

```
<library path="libnodelet_hello_world">
<class name="nodelet_hello_world/Hello" type="nodelet_hello_world::Hello"
base_class_type="nodelet::Nodelet">
      <description>
      A node to duplicate a message
      </description>
</class>
</library>
```

Moreover, we will also add the export tag in `package.xml`, along with the build and runtime dependencies (optional):

```
<export>
    <nodelet plugin="${prefix}/hello_ros.xml"/>
</export>
```

Building and running nodelets

We will make the entry for the source code files in `CMakeLists.txt`, so that we can build a `nodelet` package:

```
## Declare a cpp library
 add_library(nodelet_hello_ros
   src/hello_ros.cpp
 )

## Specify libraries to link a library or executable target against
 target_link_libraries(nodelet_hello_ros
   ${catkin_LIBRARIES}
 )
```

Consequently, we could build the package using `catkin_make` and, if the build is successful, it would generate a shared object `libnodelet_hello_world.so` file, which is indeed a plugin.

The following commands can be used to start the `nodelet manager`:

```
$ roscore
$ rosrun nodelet nodelet manager __name:=nodelet_manager
```

If `nodelet manager` runs successfully, we will see a message as shown in the following screenshot:

```
^Ckumar@kumar-Inspiron-5437:~/catkin_ws$ rosrun nodelet nodelet manager __name:=nodelet_manager
[ INFO] [1513429439.559984285]: Initializing nodelet with 4 worker threads.
```

The nodelet manager

After launching `nodelet manager`, we have to start `nodelet` by using the following command:

```
$ rosrun nodelet nodelet load nodelet_hello_world/Hello nodelet_manager
__name:=Hello1
```

Upon execution of the preceding command, `nodelet` contacts `nodelet manager` to spawn an instance of the `nodelet_hello_ros/Hello` nodelet with a name of `Hello1`. If the nodelet is instantiated successfully, we will get a message as shown here:

```
kumar@kumar-Inspiron-5437:~/catkin_ws$ rosrun nodelet nodelet load nodelet_hello_ros/Hello nodelet_manager __name:=Hello1
[ INFO] [1513436103.522021253]: Loading nodelet /Hello1 of type nodelet_hello_ros/Hello to manager nodelet_manager with the following remapping
s:
```

The running nodelet

We can look at the list of topics generated after running this nodelet, such as that shown in the following screenshot:

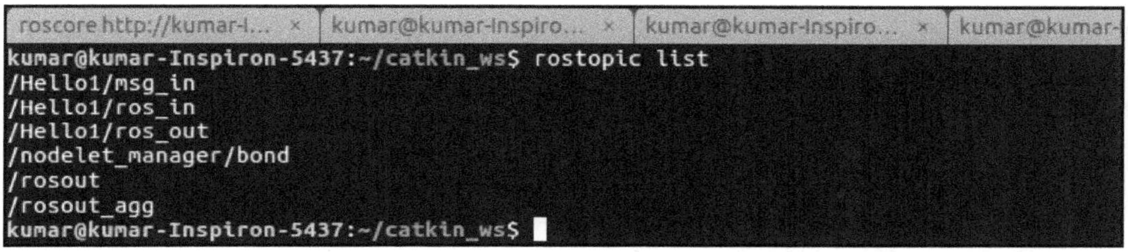

```
roscore http://kumar-i...   ×   kumar@kumar-inspiro...   ×   kumar@kumar-inspiro...   ×   kumar@kumar-
kumar@kumar-Inspiron-5437:~/catkin_ws$ rostopic list
/Hello1/msg_in
/Hello1/ros_in
/Hello1/ros_out
/nodelet_manager/bond
/rosout
/rosout_agg
kumar@kumar-Inspiron-5437:~/catkin_ws$ 
```

Nodelet topics

We can also verify the workings of certain nodelets by publishing a message string to the `/Hello1/ros_in` topic and looking for whether the same message is received at `Hello1/ros_out`:

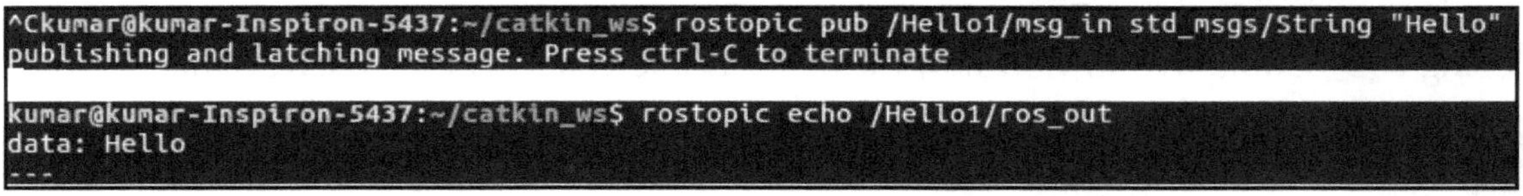

```
^Ckumar@kumar-Inspiron-5437:~/catkin_ws$ rostopic pub /Hello1/msg_in std_msgs/String "Hello"
publishing and latching message. Press ctrl-C to terminate

kumar@kumar-Inspiron-5437:~/catkin_ws$ rostopic echo /Hello1/ros_out
data: Hello
---
```

Working nodelets

Here, you will notice that a single instance of the `Hello()` class is created as a node. Moreover, we can also create multiple instances of the `Hello()` class with different node names inside this nodelet.

There's more...

As discussed previously, we could also create launch files to load more than one instance of the nodelet class. The following launch file, hello_ros.launch, will load two nodelets with the names Hello1 and Hello2:

```
<launch>
<!-- Started nodelet manager -->

  <node pkg="nodelet" type="nodelet" name="standalone_nodelet"
args="manager" output="screen"/>
<!-- Starting first nodelet -->

  <node pkg="nodelet" type="nodelet" name="Hello1" args="load
nodelet_hello_ros/Hello standalone_nodelet" output="screen">
  </node>

<!-- Starting second nodelet -->

  <node pkg="nodelet" type="nodelet" name="Hello2" args="load
nodelet_hello_ros/Hello standalone_nodelet" output="screen">
  </node>

</launch>
```

We can launch the nodelets with this launch file as follows:

```
$ roslaunch nodelet_hello_world hello_ros.launch
```

Using rqt_graph, we can view how the nodelets interconnect inside the ROS framework:

```
$rosrun rqt_gui rqt_gui
```

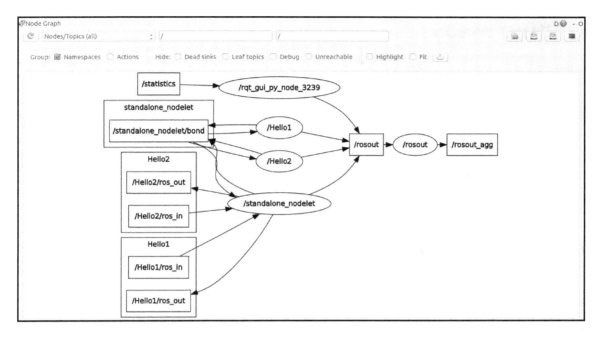

ROS computational network graph showing two instances of a nodelet

Understanding the Gazebo Framework and plugin

Gazebo is a 3D simulator framework which can accurately and efficiently simulate a wide range of robots in a complex environment. Although it is a simulator framework, Gazebo is similar to game engines but offers physics simulation at a much higher degree of consistency and integrity.

In this section, we will familiarize ourselves with the basics of Gazebo plugin development, which will be useful in the upcoming chapter Chapter 4, *ROS Visualization and Debugging Tools* where will we discuss the development of a simulator for various kinds of robots. However, understanding the Gazebo plugin requires knowledge of the Gazebo framework, which will be discussed in Chapter 6, *ROS Modelling and Simulation*; Chapter 7, *Mobile Robot in ROS*; Chapter 8, *The Robotic Arm in ROS*, and Chapter 9, *Micro Aerial Vehicle in ROS*.

Getting ready

Gazebo plugins enable us to develop several robot models, sensors, and world properties using the Gazebo framework. Similar to `pluginlib` and `nodelet`, Gazebo plugins are share libraries, which can be dynamically loaded or unloaded from the Gazebo simulator.

However, Gazebo is independent of the ROS framework, and so using plugins, we can access and extend all the components of Gazebo. As usual, we can organize the Gazebo plugins into six types:

- World
- Model
- Sensor
- System
- Visual
- GUI

Each of these plugin types is associated with different components of Gazebo.

For example, a Model plugin is attached to and controls a specific model in Gazebo such as a holonomic or non-holonomic robot model. Similarly, a World plugin is attached to a world such as a structure or unstructured environment, and a Sensor plugin is attached to a specific sensor. Furthermore, the System plugin is specified on the command line, which is loaded first during Gazebo startup and provide users control over the startup process. Moreover, Visual and GUI plugins are related to appearance and are attached to the rendering engine.

A plugin type should be selected based on the desired functionality.

For example, you should use a World plugin to control world properties, such as the physics engine, ambient lighting, and so on. Similarly, you should use a Model plugin to control joints and the state of a model. Furthermore, you should use a Sensor plugin to acquire sensor information and control sensor properties.

How to do it...

Before starting development with Gazebo plugins, we have to install the dependent packages. This is because we are working with ROS Kinetic which has Gazebo version 7.x by default. Therefore, we have to install Gazebo's development package, `libgazebo7-dev`:

```
$sudo apt-get install libgazebo7-dev
```

The Gazebo plugins are independent of ROS and we don't need the ROS libraries to build the plugin.

Hello World plugin

Plugins are designed to be simple. Here, we will discuss a bare bones World plugin which contains a class with a few member functions.

First, we will make a directory and a .cc file for the new plugin:

```
$ mkdir ~/gazebo_plugin_tutorial
$ cd ~/gazebo_plugin_tutorial
$ gedit hello_world.cc
```

You can also use the existing package from the chapter3_tutorials/ gazebo_plugin_tutorial file on GitHub.

Next, we will add the following code to hello_world.cc:

```
#include <gazebo/gazebo.hh>
namespace gazebo
{
  class WorldPluginTutorial : public WorldPlugin
  {
    public: WorldPluginTutorial() : WorldPlugin()
            {
              printf("Gazebo Says: Hello World!n");
            }
    public: void Load(physics::WorldPtr _world, sdf::ElementPtr _sdf)
            {
            }
  };
  GZ_REGISTER_WORLD_PLUGIN(WorldPluginTutorial)
}
```

In the preceding code, the gazebo/gazebo.hh file includes a core set of basic Gazebo functions. Since this is a case by case basis and type of plugin, it should include gazebo/physics/physics.hh, gazebo/rendering/rendering.hh, or gazebo/sensors/sensors.hh. Moreover, all of the plugins must be in the gazebo namespace and must inherit from a plugin type, which in this case is the WorldPlugin class. Here, one of the required functions is Load(), which receives an SDF element that contains the elements and attributes that are specified in the loaded SDF file. We will discuss more about the SDF file in the upcoming chapter, Chapter 4, *ROS Visualization and Debugging Tools* while we simulate the design of the robot model and its world.

Finally, the plugin must be registered with the simulator using the `GZ_REGISTER_WORLD_PLUGIN` macro. The only parameter to this macro is the name of the plugin class. There are matching register macros for each plugin type, which are as follows: `GZ_REGISTER_MODEL_PLUGIN`, `GZ_REGISTER_SENSOR_PLUGIN`, `GZ_REGISTER_GUI_PLUGIN`, `GZ_REGISTER_SYSTEM_PLUGIN`, and `GZ_REGISTER_VISUAL_PLUGIN`. They should be used on a case by case basis.

Compiling the plugin

We should make sure that Gazebo has been properly installed. To compile the preceding plugin, we will create `~/gazebo_plugin_tutorial/CMakeLists.txt` and add the rule for compilation:

```
cmake_minimum_required(VERSION 2.8 FATAL_ERROR)
find_package(gazebo REQUIRED)
include_directories(${GAZEBO_INCLUDE_DIRS})
link_directories(${GAZEBO_LIBRARY_DIRS})
list(APPEND CMAKE_CXX_FLAGS "${GAZEBO_CXX_FLAGS}")

add_library(hello_world SHARED hello_world.cc)
target_link_libraries(hello_world ${GAZEBO_LIBRARIES})
set(CMAKE_CXX_FLAGS "${CMAKE_CXX_FLAGS} ${GAZEBO_CXX_FLAGS}")
```

Next, we will create the `build` directory and compile the code:

```
$ mkdir ~/gazebo_plugin_tutorial/build
$ cd ~/gazebo_plugin_tutorial/build
$ cmake ..
$ make
```

The successful compilation will generate a shared library, `~/gazebo_plugin_tutorial/build/libhello_world.so`, that can be inserted in a Gazebo simulation. Lastly, we have to add a library path to the `GAZEBO_PLUGIN_PATH`:

```
$ export
GAZEBO_PLUGIN_PATH=${GAZEBO_PLUGIN_PATH}:~/gazebo_plugin_tutorial/build
```

This only adds the path for the current shell. If we want to use our plugin for every new terminal we open, we have to append the preceding line to the `~/.bashrc` file.

Using a plugin

We already have a plugin that compiles as a shared library, and so we could attach it to a world model in an SDF file (refer to the SDF documentation for more information). During startup, Gazebo parses the SDF file, locates the plugin, and loads the code.

Next, we will create a `world` file called `~/gazebo_plugin_tutorial/hello.world` and copy the following code into it:

```xml
<?xml version="1.0"?>
<sdf version="1.4">
  <world name="default">
    <plugin name="hello_world" filename="libhello_world.so"/>
  </world>
</sdf>
```

Finally, we will pass it to `gzserver` on startup:

```
$ gzserver ~/gazebo_plugin_tutorial/hello.world --verbose
```

We will see output similar to this:

```
kumar@kumar-Inspiron-5437:~/gazebo_plugin_tutorial$ export GAZEBO_PLUGIN_PATH=${GAZEBO_PLUGIN_PATH}:~/gazebo_plugin_tutorial/build
kumar@kumar-Inspiron-5437:~/gazebo_plugin_tutorial$ gzserver ~/gazebo_plugin_tutorial/hello.world --verbose
Gazebo multi-robot simulator, version 8.1.1
Copyright (C) 2012 Open Source Robotics Foundation.
Released under the Apache 2 License.
http://gazebosim.org

[Msg] Waiting for master.
[Msg] Connected to gazebo master @ http://127.0.0.1:11345
[Msg] Publicized address: 192.168.1.2
Hello World!
```

Gazebo plugin

Understanding the ROS transform frame (TF)

The ROS TF library has been developed to provide a standard method in order to keep track of coordinate frames and transform data within the entire system so that individual component users can be confident about the consistency of their data in a particular coordinate frame without requiring knowledge about all the other coordinate frames in the system and their associations.

Getting ready

In this section, we will discuss TF and its uses in robotics application development. We will learn about some of the power TF has in a multi-robot example using turtlesim and also introduce you to TF visualization and debugging tools such as `tf_echo`, `view_frames`, `rqt_tf_tree`, and `rviz`.

First of all, we will install the required packages from the ROS repository:

```
$ sudo apt-get install ros-kinetc-ros-tutorials ros-kinetic-geometry-
tutorials ros-kinetic-rviz ros-kinetic-rosbash ros-kinetic-rqt-tf-tree
```

After installing the required packages, we will start the `demo` example:

```
$ roslaunch turtle_tf turtle_tf_demo.launch
```

Once the `turtlesim` demo is started, we will drive the center turtle around in `turtlesim` using the keyboard arrow keys.

We have to select the `roslaunch` terminal window or select the **Always Top** option for the Terminal so that our keystrokes will be captured to drive the turtle.

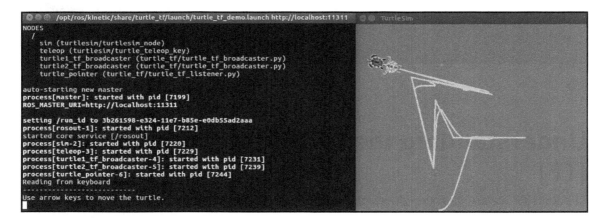

Showing turtle2 following turtle1 using TF

We can observe that one turtle will continuously follow the turtle we are driving. In this demo application, the ROS TF library is used to create three coordinate frames: a world frame, a `turtle1` frame, and a `turtle2` frame, and to create a TF broadcaster to publish the coordinate frames of the first turtle and a TF listener to compute the difference between the first and follower turtle frames, as well as drive the second turtle to follow the first.

We will now discuss a few ROS tools which are used to visualize and debug the TF transformation.

Using view_frames

The `view_frames` tool creates a diagram of the frames being broadcast by TF over ROS:

```
$ rosrun tf view_frames
$ evince frames.pdf
```

The diagram is shown as follows:

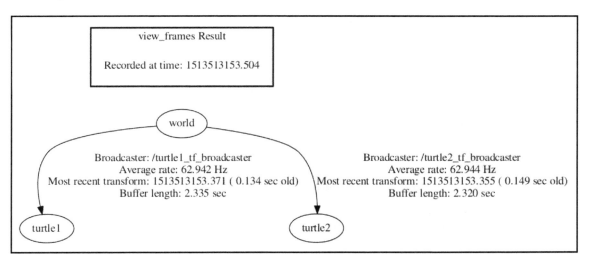

Showing TF relation between turtle1 and turtle2

Here, you may have noticed that three frames are broadcast by TF—the world, `turtle1`, and `turtle2`, where the world frame is the parent of the `turtle1` and `turtle2` frames.

Using rqt_tf_tree

The `rqt_tf_tree` tool enables the real-time visualization of the tree of frames being broadcast over ROS:

```
$ rosrun rqt_tf_tree rqt_tf_tree
```

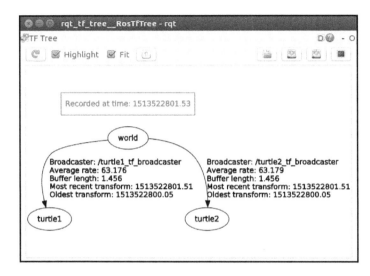

TF demo TF tree

Using tf_echo

The `tf_echo` tool reports the transformation between any two frames broadcast over ROS:

```
$ rosrun tf tf_echo [reference_frame] [target_frame]
$ rosrun tf tf_echo turtle1 turtle2
```

You may have noticed that the transform is displayed as the `tf_echo` listener receives the frame's broadcast over ROS:

```
kumar@kumar-Inspiron-5437:~$ rosrun tf tf_echo turtle1 turtle2
At time 1513513192.139
- Translation: [0.000, 0.000, 0.000]
- Rotation: in Quaternion [-0.000, -0.000, 0.135, 0.991]
            in RPY (radian) [-0.000, -0.000, 0.271]
            in RPY (degree) [-0.000, -0.000, 15.540]
At time 1513513192.876
- Translation: [0.000, 0.000, 0.000]
- Rotation: in Quaternion [-0.000, -0.000, 0.135, 0.991]
            in RPY (radian) [-0.000, -0.000, 0.271]
            in RPY (degree) [-0.000, -0.000, 15.540]
```

TF demoing echo frames

Using RViz and TF

`rviz` is a graphical 3D visualization tool that is useful for viewing the association between TF frames within the ROS system:

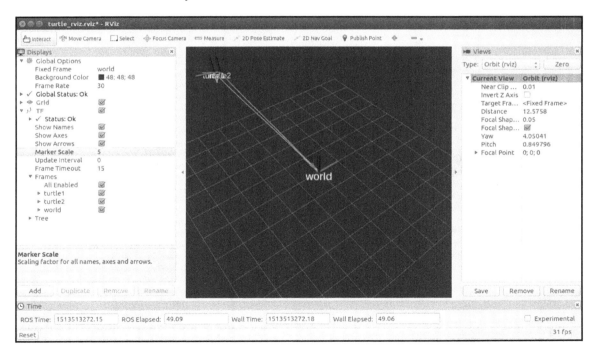

TF demo RViz

Here, you may have noticed that the frames are being broadcast by TF in the sidebar and as we drive the turtle around, and that we can observe the frame moving in `rviz`.

How to do it...

In this subsection, we will learn how to write the code to reproduce the demo application discussed previously.

Writing a TF broadcaster

First, we will create a `ros` package in our workspace or copy the `chapter3_tutorials/tf_tutorials` file from GitHub:

```
$ catkin_create_pkg tf_tutorials tf roscpp rospy turtlesim
```

Next, we will create the `tf_tutorials/src/turtle_tf_broadcaster.cpp` file and add the source code:

```cpp
#include <ros/ros.h>
#include <tf/transform_broadcaster.h>
#include <turtlesim/Pose.h>

std::string turtle_name;


void poseCallback(const turtlesim::PoseConstPtr& msg){
  static tf::TransformBroadcaster br;
  tf::Transform transform;
  transform.setOrigin( tf::Vector3(msg->x, msg->y, 0.0) );
  tf::Quaternion q;
  q.setRPY(0, 0, msg->theta);
  transform.setRotation(q);
  br.sendTransform(tf::StampedTransform(transform, ros::Time::now(),
"world", turtle_name));
}

int main(int argc, char** argv){
  ros::init(argc, argv, "tf_broadcaster");
  if (argc != 2){ROS_ERROR("need turtle name as argument"); return -1;};
  turtle_name = argv[1];

  ros::NodeHandle node;
  ros::Subscriber sub = node.subscribe(turtle_name+"/pose", 10,
&poseCallback);
```

```
    ros::spin();
    return 0;
};
```

Here, we will create a `TransformBroadcaster` object that will be used to send the transformations over the ROS communication network. And, we will also create a `Transform` object and copy the information from the 2D turtle pose into the 3D transformation:

```
br.sendTransform(tf::StampedTransform(transform, ros::Time::now(), "world",
turtle_name));
```

In the preceding code snippet, the real work is done where `Transform` with `TransformBroadcaster` requires four arguments:

- First, we will pass in transform itself, which was created in the preceding code snippet.
- Next, we provide a timestamp to the transform being published. We'll just stamp it with the current time, `ros::Time::now()`.
- Then, we will provide the name of the parent frame of the link we have created, in this case, "world".
- Finally, we will provide the name of the child frame of the link we have created, in this case, the name of the turtle itself.

 `sendTransform` and `StampedTransform` have opposite ordering of parent and child.

Writing a tf listener

In the previous subsection, we created a TF broadcaster to publish the pose of a turtle to TF. In this subsection, we will create a TF listener that will use that TF.

We will create a `tf_tutorials/src/ turtle_tf_listener.cpp` file and add the following source code to it:

```
#include <ros/ros.h>
#include <tf/transform_listener.h>
#include <geometry_msgs/Twist.h>
#include <turtlesim/Spawn.h>

int main(int argc, char** argv){
```

```
ros::init(argc, argv, "tf_listener");

ros::NodeHandle node;

ros::service::waitForService("spawn");
ros::ServiceClient add_turtle =
  node.serviceClient<turtlesim::Spawn>("spawn");
turtlesim::Spawn srv;
add_turtle.call(srv);

ros::Publisher turtle_vel =
  node.advertise<geometry_msgs::Twist>("turtle2/cmd_vel", 10);

tf::TransformListener listener;

ros::Rate rate(10.0);
while (node.ok()){
  tf::StampedTransform transform;
  try{
    listener.lookupTransform("/turtle2", "/turtle1",
                             ros::Time(0), transform);
  }
  catch (tf::TransformException &ex) {
    ROS_ERROR("%s",ex.what());
    ros::Duration(1.0).sleep();
    continue;
  }

  geometry_msgs::Twist vel_msg;
  vel_msg.angular.z = 4.0 * atan2(transform.getOrigin().y(),
                                  transform.getOrigin().x());
  vel_msg.linear.x = 0.5 * sqrt(pow(transform.getOrigin().x(), 2) +
                                pow(transform.getOrigin().y(), 2));
  turtle_vel.publish(vel_msg);

  rate.sleep();
}
return 0;
};
```

Here, we will create a `TransformListener` object which will start receiving TF transformations over the ROS communication network, and will buffer them for up to 10 seconds by default:

```
try{
    listener.lookupTransform("/turtle2", "/turtle1",
                             ros::Time(0), transform);
}
```

In the preceding line of source code, we queried the listener for a specific transformation that will take four arguments:

- In the first two arguments, we will transform from the `/turtle1` frame to the `/turtle2` frame.
- In the next argument, the time at which the transformation is required, providing `ros::Time(0)` will set up the latest available transform.
- In the final argument, we will use the object to store the resulting transformation.

Compiling and running the TF

In the preceding subsection, we had to learn to develop the code for the TF demo application. To compile this, we will have to make an entry in `CMakeLists.txt` about source files and compilation rules:

```
add_executable(turtle_tf_broadcaster src/turtle_tf_broadcaster.cpp)
target_link_libraries(turtle_tf_broadcaster ${catkin_LIBRARIES})

add_executable(turtle_tf_listener src/turtle_tf_listener.cpp)
target_link_libraries(turtle_tf_listener ${catkin_LIBRARIES})
```

Now, we can build the package from the top folder of the workspace using `catkin_make`. If everything goes well, we will have binary files called `turtle_tf_broadcaster` and `turtle_tf_listner` in our `devel/lib/` folder. To start the application, we will create a launch file called `launch/start_demo.launch` and add the startup, configuration, and information there:

```
<launch>
    <!-- Turtlesim Node-->
    <node pkg="turtlesim" type="turtlesim_node" name="sim"/>

    <node pkg="turtlesim" type="turtle_teleop_key" name="teleop"
output="screen"/>
    <!-- Axes -->
```

```
<param name="scale_linear" value="2" type="double"/>
<param name="scale_angular" value="2" type="double"/>

<node pkg="tf_tutorials" type="turtle_tf_broadcaster"
      args="/turtle1" name="turtle1_tf_broadcaster" />
<node pkg="tf_tutorials" type="turtle_tf_broadcaster"
      args="/turtle2" name="turtle2_tf_broadcaster" />
<node pkg="tf_tutorials" type="turtle_tf_listener"
      name="listener" />
</launch>
```

Finally, we can launch the application and verify the result using the TF visualization and debugging tools which we discussed in the previous subsection:

```
$ roslaunch tf_tutorials start_demo.launch
```

Understanding the ROS Visualization tool (RViz) and its plugins

The RViz tool is recommended and it is the standard 3D visualization and debugging tool in the ROS system. Almost all kinds of data from sensors and various states of robots and their correspondence with 3D worlds can be viewed through this tool. Moreover, RViz will be installed along with the ROS desktop's full installation.

Getting ready

In this section, we will launch rviz and discuss the basic components of RViz:

```
$rosrun rviz rviz
```

We assume that `roscore` is running:

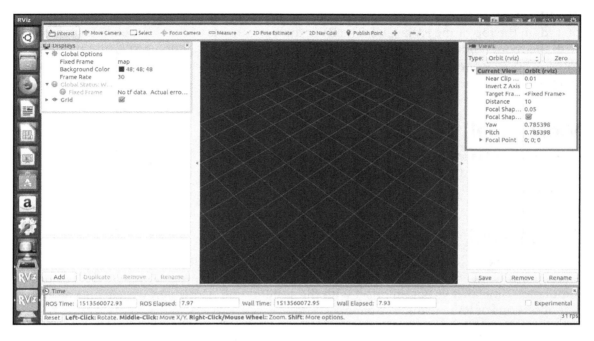

Rviz

In the preceding screenshot, the RViz GUI is marked and we will discuss the uses of each section.

Display panel

The panel on the left side of the RViz is known as the **Displays** panel, which contains a list of display plugins for RViz and their properties. These display plugins are used to visualize different kind of ROS messages, mostly different types of sensor data in the RViz 3D viewport and robot model state information and their correspondence, for example, viewing images from a camera, viewing a 3D point cloud from a laser scan, the robot model, TF, and many more. The required plugins can be added by pressing the Add button on the left panel. Moreover, we can also develop specific display plugins and add them there.

RViz toolbar

The RViz toolbar tool is used to manipulate the 3D viewport of the RViz display, which is placed on the top of the RViz GUI. The RViz toolbar consists of a set of tools for interacting with the robot model, modifying the camera view, setting navigation goals, and providing robot 2D pose estimations. Moreover, we can add custom tools on the toolbar in the form of plugins.

View panel

The view panel is available on the right-hand side of the RViz GUI, which lets you save different views of the 3D viewport and switch to each view by loading the saved configuration.

Time panel

The **Time** panel is located at the bottom of the RViz GUI, and it displays the elapsed simulator time. It is most useful when there is a simulator running along with RViz since simulation time always has an offset with the system time. Moreover, we can also reset the initial settings of RViz using this panel.

We can refer to the detailed documentation for developing plugins for RViz at `http://docs.ros.org/jade/api/rviz_plugin_tutorials/html/index.html`.

This is where the `rviz_plugin_tutorials` package builds a plugin library for RViz containing two main classes—`ImuDisplay` and `TeleopPanel`. Will discuss the `ImuDisplay` plugin in the following subsection.

Developing an RViz plugin for IMU Display

In this subsection, we will learn how to write a simple Display plugin for RViz. Since RViz does not currently have a way to `display sensor_msgs/Imu` messages directly, we will implement a subclass of `rviz::Display` to do so.

How to do it...

The source code for the `rviz_plugin_tutorials` package can be downloaded from the `chapter3_tutorials/rviz_plugin_tutorials/` file, which contains the code for ImuDisplay in these files: `src/imu_display.h`, `src/imu_display.cpp`, `src/imu_visual.h`, and `src/imu_visual.cpp`.

Alternatively, we can create a package with the `ImuDisplay` plugin and add the source code in the respective files:

```
$ catkin_create_pkg rviz_plugin_tutorials roscpp rviz std_msgs
```

The open source code is self-explanatory.

Exporting the plugin

We will create a `plugin_description.xml` file so that the plugin can be found and understood by other ROS packages in the system, as we have done in the case of plugins in the previous subsection:

```
<export>
    <rviz plugin="${prefix}/plugin_description.xml"/>
</export>
```

The `plugin_description.xml` file has the following entry at least:

```
<library path="lib/librviz_plugin_tutorials">
  <class name="rviz_plugin_tutorials/Imu"
         type="rviz_plugin_tutorials::ImuDisplay"
         base_class_type="rviz::Display">
    <description>
      Displays direction and scale of accelerations from sensor_msgs/Imu
messages.
    </description>
    <message_type>sensor_msgs/Imu</message_type>
  </class>
</library>
```

Building and working with the plugin

We can build the plugin by invoking `catkin_make` from the top level folder as usual. Once the RViz plugin is compiled and exported, we will simply run `rviz` normally and `rviz` will use `pluginlib` to find all the plugins exported to it:

```
$ rosrun rviz rviz
```

The output is shown in the following screenshot:

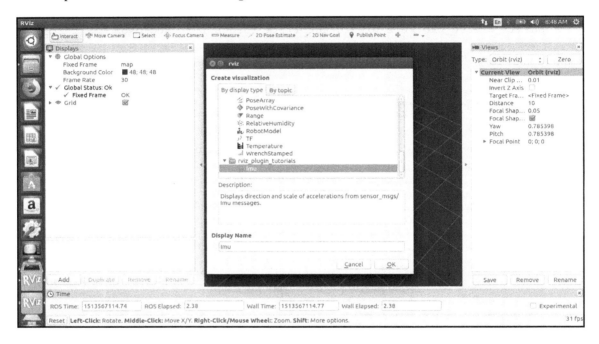

RViz adding the IMU plugin

We will add `ImuDisplay` by clicking the **Add** button at the bottom of the **Displays** panel, as shown in the preceding screenshot. Since we are working in a simulated environment and do not have an IMU or another source of `sensor_msgs/Imu` messages, we can even verify the plugin with a Python script which will simulate the sensor outputs on the `/test_imu` topic, which must be set in the display panel for `Imudisplay` in the RViz GUI:

```
$ python scripts/send_test_msgs.py
```

The Python script publishes both the IMU messages and a TF frame which will be visualized in the RViz GUI:

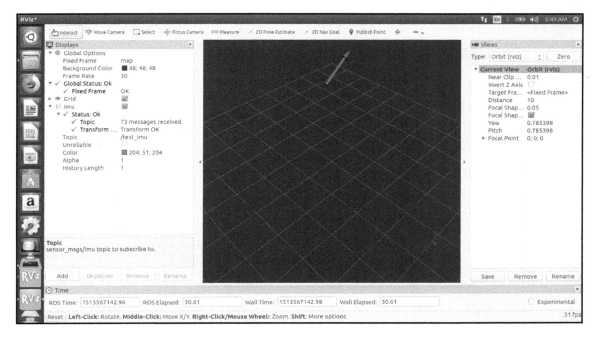

RViz running the IMU plugin

We can refer to the details for developing the `TeleopPanel` and `PlantFlagTool` plugins from the following link `http://docs.ros.org/kinetic/api/rviz_plugin_tutorials/html/index.html`.

4

ROS Visualization and Debugging Tools

In this chapter, we will discuss the following recipes:

- Debugging and profiling ROS nodes
- Logging and visualizing ROS messages
- Inspecting and diagnosing the ROS system
- Visualizing and plotting scalar data
- Visualizing non-scalar data – 2D/3D images
- Recording and playing back ROS topics

Introduction

In the previous chapter, we looked into the advanced concepts of the ROS architecture and its functionality, which started with the parameter server, `actionlib`, `pluginlib`, and `nodelet`. Moreover, we also looked at the Gazebo framework and plugin for simulation, along with the ROS TF (transform frame). However, we also started discussing the **ROS visualization tool** (**RViz**) and its plugins, which are used for debugging and monitoring.

In this chapter, we will discuss certain concepts and tools used in visualization and debugging. First, we will learn how to debug and profile a ROS node with the GDB debugger and Valgrind.

Consequently, we will look into how to log and visualize ROS messages and observe the ROS computational graph with different severity levels, names, conditions, and throttling options using `rqt`.

Similarly, in this chapter, we will learn how to plot scalar data in a discrete time series, visualize images from a video stream, and represent different types of vector data in 3D using RViz.

Debugging and profiling ROS nodes

ROS nodes can be debugged in a similar way to conventional programs since they run as a process with the associated `pid` in the Linux runtime environment. Therefore, we can debug them using standard tools, such as `gdb`, check for memory leaks with `memcheck`, and profile for performance using `valgrind` or both.

In the following sections, we will learn how to configure these tools in the ROS framework. Next, we will look into how to add logging messages to the source code of programs in order to make them observable so that we can diagnose basic problems in runtime without the need to debug the executable binaries. Later, we will discuss ROS introspection tools, which help in easily detecting broken connections between nodes.

Getting ready

First, we are going to create each example program in the ROS package in our workspace individually. Alternatively, we could copy the `chapter4_tutorials` package from GitHub at `https://github.com/kbipin/Robot-Operating-System-Cookbook`.

We can debug a ROS C/C++ node with the `gdb` debugger by using the location of the node executable and running the node executable inside `gdb` with the following command:

```
$ cd devel/lib/chapter4_tutorials
$ gdb program1
```

 `roscore` must be running before we start our node inside `gdb` since it will need the master server to be running. We have to use the `catkin_make -DCMAKE_BUILD_TYPE=Debug` command to build the ROS packages with the debug symbol.

How to do it...

1. Attach the `gdb` debugger with the `launch` file as follows:

```
<launch>
  <node pkg=""chapter4_tutorials"" type=""program1""
name=""program1"" output=""screen""
  launch-prefix=""xterm -e gdb --args""/>
</launch>
```

2. Start the ROS node within `gdb` debugger using the `launch` command:

```
$ roslaunch chapter4_tutorials program1_gdb.launch
```

3. Create a new `xterm` terminal with the node attached to `gdb`, as shown in the following screenshot.

Moreover, we can set breakpoints if needed and then press the *c* or *r* key to run the node and debug it. The source code can be listed by using the `l` command:

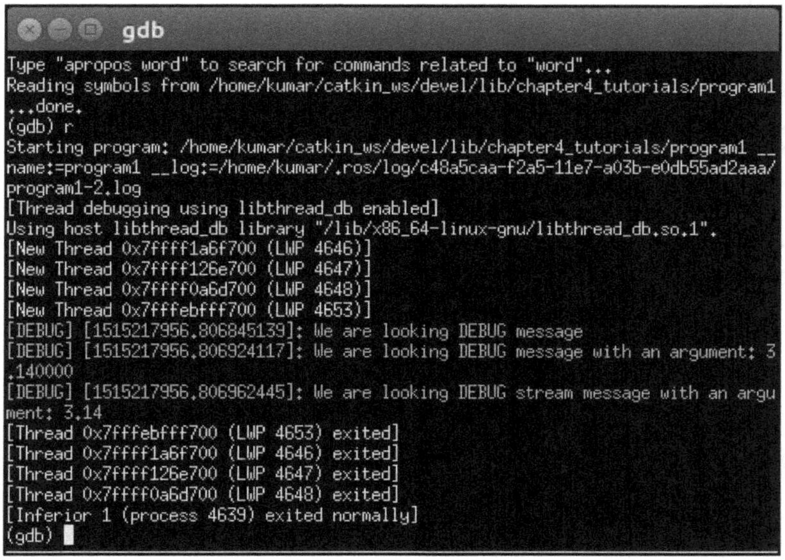

ROS node with gdb debugger

4. Try debugging `python` in a similar way:

```
$ gdb python
$ run program.py
```

5. Again, attach the `gdb` debugger with the `launch` file as follows:

```
<launch>
  <node pkg=""chapter4_tutorials"" type=""program.py""
name=""program.py"" output=""screen""
  launch-prefix=""xterm -e gdb --args""/>
</launch>
```

Actually, ROS nodes are typically executable, but particular techniques are required to enable core dumps, which can be used in a `gdb` session later. First of all, we have to set an unlimited core size as follows:

```
$ ulimit -c unlimited
```

This is also required for any regular executable and not only for ROS nodes. Next, in order to create core dumps with the name and path `$ROS_HOME/core.PID`, we will run the following command for setting the kernel configuration using the `proc` filesystem:

```
$ echo 1 | sudo tee /proc/sys/kernel/core_uses_pid
```

6. Start the ROS node within the `gdb` debugger using the `launch` command:

```
$ roslaunch chapter4_tutorials program1_dump.launch
```

7. Load the symbol file in the `gdb` session upon executing the `file` command as follows:

```
gdb> file
/home/kumar/catkin_ws/devel/lib/chapter4_tutorials/program1_dum
p
```

This will create a new `xterm` terminal with the node attached to `gdb`, as shown in the following screenshot. We can observe the crash message and also call the stack using the `bt` command. We can also call the `core.pid.dump` at `~/.ros` directory by default:

ROS node core dump

Furthermore, we can use a similar attribute to attach the node to diagnosis tools such as Valgrind. This allows us to detect memory leaks using `memcheck` and perform profiling analysis using `callgrind` (refer to `http://valgrind.org` for more information):

```
<launch>
  <!-- Program 1 with Memory Profiler valgrind -->
  <node pkg=""chapter4_tutorials"" type=""program1_mem""
name=""program1_mem"" output=""screen"" launch-prefix=""valgrind""/>
</launch>
```

We will launch the example program to observe the memory leaks that have been introduced into the source code deliberately.

The Valgrind tools output the detected memory leak, as shown in the following screenshot:

```
$ roslaunch chapter4_tutorials program1_valgrind.launch
```

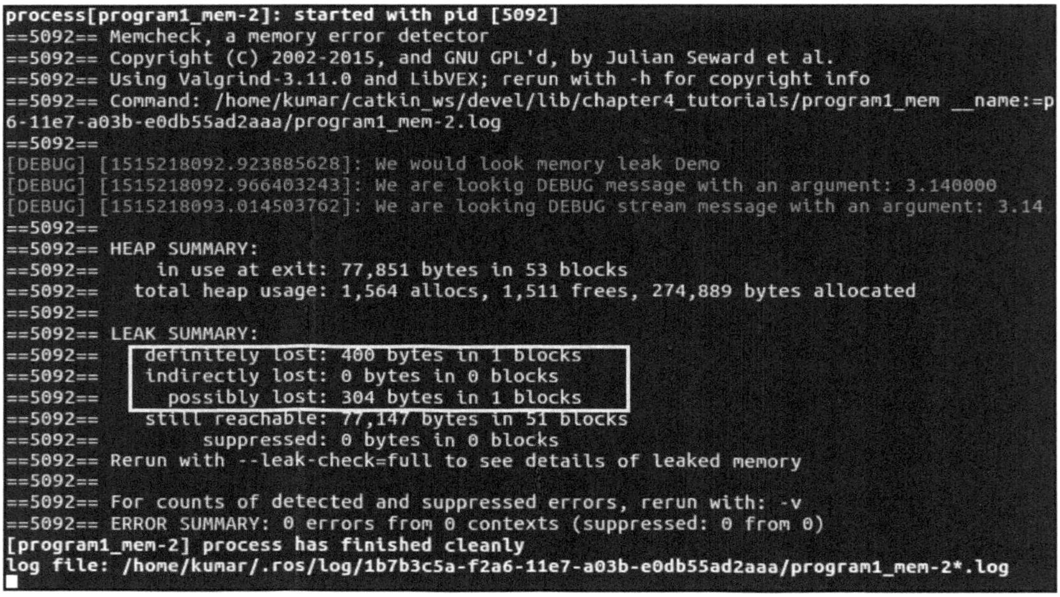

```
process[program1_mem-2]: started with pid [5092]
==5092== Memcheck, a memory error detector
==5092== Copyright (C) 2002-2015, and GNU GPL'd, by Julian Seward et al.
==5092== Using Valgrind-3.11.0 and LibVEX; rerun with -h for copyright info
==5092== Command: /home/kumar/catkin_ws/devel/lib/chapter4_tutorials/program1_mem __name:=p
6-11e7-a03b-e0db55ad2aaa/program1_mem-2.log
==5092==
[DEBUG] [1515218092.923885628]: We would look memory leak Demo
[DEBUG] [1515218092.966403243]: We are lookig DEBUG message with an argument: 3.140000
[DEBUG] [1515218093.014503762]: We are looking DEBUG stream message with an argument: 3.14
==5092==
==5092== HEAP SUMMARY:
==5092==     in use at exit: 77,851 bytes in 53 blocks
==5092==   total heap usage: 1,564 allocs, 1,511 frees, 274,889 bytes allocated
==5092==
==5092== LEAK SUMMARY:
==5092==    definitely lost: 400 bytes in 1 blocks
==5092==    indirectly lost: 0 bytes in 0 blocks
==5092==      possibly lost: 304 bytes in 1 blocks
==5092==    still reachable: 77,147 bytes in 51 blocks
==5092==         suppressed: 0 bytes in 0 blocks
==5092== Rerun with --leak-check=full to see details of leaked memory
==5092==
==5092== For counts of detected and suppressed errors, rerun with: -v
==5092== ERROR SUMMARY: 0 errors from 0 contexts (suppressed: 0 from 0)
[program1_mem-2] process has finished cleanly
log file: /home/kumar/.ros/log/1b7b3c5a-f2a6-11e7-a03b-e0db55ad2aaa/program1_mem-2*.log
```

ROS node memory leak shown by Valgrind

Logging and visualizing ROS messages

In software development, it is always good practice to include messages that indicate what the program is doing with clarity without compromising the efficiency of the software. The ROS framework provides a set of APIs to enable these features, which is built on top of log4cxx, a log4j logger library.

In short, we have several levels of messages, and all of them have a negligible footprint on performance and can be masked by the current verbosity level during compile time or even in runtime. Moreover, they are fully integrated with other ROS tools for visualization.

Getting ready

ROS has a great number of functions and macros to display log messages which support various verbosity levels, conditions, STL streams, throttling, and other special features. We are going to learn about these in this section.

We will start with a simple information message that prints the messages onto the console:

```
ROS_INFO( ""ROS INFO message."" );
```

As a result of running a program with the preceding message, the following outputs are shown on the screen:

```
[INFO] [1456880231.839068150]: ROS INFO message.
```

As we can see, all of the messages are printed with their verbosity and current timestamp followed by the actual message, which is both within square brackets. Moreover, this function also allows parameter arguments, just like the C `printf` function does. For example, in the following code, we can print the value of a floating point number into the variable, too:

```
const double val = 3.14; ROS_INFO( ""ROS INFO message with argument: %f"",
val );
```

Additionally, C++ STL streams are also supported with `*_STREAM` functions:

```
ROS_INFO_STREAM( ""ROS INFO stream message with argument: "" << val);
```

It does not specify any stream since the API takes care of that by redirecting to `cout` or `cerr`, a file, or both.

Refer to `chapter4_tutorials/src/program2.cpp` for more details.

ROS supports the following verbosity levels for logging:

- `DEBUG`: Used for debugging and testing purposes only; this information should not be displayed in a deployed application in the production system
- `INFO`: A standard message to notify what the node is doing
- `WARN`: Used for providing a warning that something might be wrong or abnormal, but that the application can still run indifferently
- `ERROR`: Indicates errors; however, the node can still recover from anomalies, but certain behaviors will be expected

- `FATAL`: Reports errors that prevent or prohibit the node from functioning

ROS uses these verbosity levels to filter the messages printed to console by a particular node. By default, only messages of INFO or higher levels are shown. There are two primary ways to set the verbosity levels. One is by using compile time and another is by using runtime, by using the configuration file or by using the `rqt_console` and `rqt_logger_level` tools.

How it works...

Setting the logging level at compile time requires code modification, which is not recommended for most cases; however, it removes the overhead of all the logging functions below a given level. Also, ROS provides ROSCONSOLE_MIN_SEVERITY as a macro variable, which must be set to the minimum desired severity level. These macros are as follows:

- ROSCONSOLE_SEVERITY_DEBUG
- ROSCONSOLE_SEVERITY_INFO
- ROSCONSOLE_SEVERITY_WARN
- ROSCONSOLE_SEVERITY_ERROR
- ROSCONSOLE_SEVERITY_FATAL
- ROSCONSOLE_SEVERITY_NONE

For example, if the desired logging level is an ERROR or higher message, we should define this in the source code as follows:

```
#define ROSCONSOLE_MIN_SEVERITY ROSCONSOLE_SEVERITY_ERROR
```

On the other hand, ROSCONSOLE_MIN_SEVERITY could be defined for all nodes in a package by setting this macro in CMakeLists.txt as follows:

```
add_definitions(-DROSCONSOLE_MIN_SEVERITY =ROSCONSOLE_SEVERITY_ERROR)
```

Alternatively, ROS provides more flexible methods of setting the minimum logging level in a configuration file at runtime. We will create a `config` folder and a file named `chapter4_tutorials.config` there, with the following content:

```
log4j.logger.ros.chapter4_tutorials=ERROR
```

We can also refer to the implementation at `chapter4_tutorials/config/chapter4_tutorials.config` at GitHub.

Then, we must set the `ROSCONSOLE_CONFIG_FILE` environment variable to assign the `config` file, which we created previously. This can be easily done in the launch file as follows (refer to `chapter4_tutorials/launch/program1.launch` for more information):

```
<launch>
<!-- Logger config --> <env name=""ROSCONSOLE_CONFIG_FILE"" value=""$(find
chapter4_tutorials)/config/chapter4_tutorials.config""/>    <!-- Program 1
--> <node pkg=""chapter4_tutorials"" type=""program1"" name=""program1""
output=""screen""/>
</launch>
```

Here, the environment variable takes the configuration file defined previously, which contains the logging level specification for each named logger. By default, ROS assigns names after the node's name to the messages. However, in complex nodes, we could provide a name to those messages based upon a given module or its functionality. This can be done with the `ROS_<LEVEL>[_STREAM]_NAMED` functions (refer to `chapter4_tutorials/src/program2.cpp` for more information):

```
ROS_INFO_STREAM_NAMED( ""named_msg"", ""ROS INFO named message."" );
```

Moreover, with the named message, we can define different initial logging levels for each named message using the configuration file:

```
log4j.logger.ros.chapter4_tutorials.named_msg=ERROR
```

Furthermore, ROS provides conditional messages that are only printed when a given condition is satisfied. These can be used using the `ROS_<LEVEL>[_STREAM]_COND[_NAMED]` functions (refer to `chapter4_tutorials/src/program2.cpp` for more examples and combinations):

```
/* Conditional messages: */ ROS_INFO_STREAM_COND(val < 0., ""ROS
conditional INFO stream message; val ("" << val << "") < 0"");
```

There are filtered messages that are inherently similar to conditional messages. However they are enabled to specify a user-defined filter that extends `ros::console::FilterBase`. The following example can be found in `program2.cpp`:

```
/* Filtered messages: */ struct ROSLowerFilter : public
ros::console::FilterBase { ROSLowerFilter( const double& val ) : value( val
) {} inline virtual bool isEnabled() { return value < 0.; } double value;
}; ROSLowerFilter filter_lower(val); ROS_INFO_STREAM_FILTER( &filter_lower,
""ROS filter INFO stream message; val ("" << val << "") < 0"" );
```

We can also control how many times a given message should be shown using
ROS_<LEVEL>[_STREAM]_ONCE[_NAMED]:

```
/* Once messages: */ for( int i = 0; i < 10; ++i ) { ROS_INFO_STREAM_ONCE(
""ROS once INFO stream message; i = "" << i ); }
```

However, it is usually better to show the message with a certain frequency. This is possible
by ROS_<LEVEL>[_STREAM]_THROTTLE[_NAMED] throttle messages. Additionally, they
have a first argument, which is a period in seconds, which defines the frequency of
printing:

```
/* Throttle messages: */ for( int i = 0; i < 10; ++i ) {
ROS_INFO_STREAM_THROTTLE( 2, ""ROS throttle INFO stream message; i = "" <<
i ); ros::Duration(1).sleep(); }
```

Finally, the named, conditional, and once or throttle messages can be used together with all
the available levels. nodelet also have support in terms of logging messages, where all the
macros are discussed until they are valid, but instead using OS_*, we have to use
NODELET_*. Basically, these macros will only compile inside *nodelets*.

There's more...

ROS Kinetic provides two major GUI tools to manage the logging message
rqt_logger_level, to set the logging level of the nodes and rqt_console, which is used
to visualize, filter, and analyze the logging messages. In this section, we will learn how to
use these tools:

```
$ roslaunch chapter4_tutorials program3.launch
```

The following command will start the program3 example node. Alternatively, we could
run the rqt_console tool:

```
$ rosrun rqt_console rqt_console
```

The following window will be open:

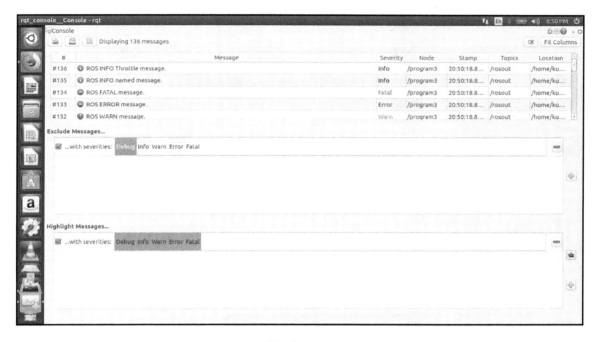

GUI tool rqt_console

In the preceding screenshot, we can see the logging message in the `rqt_console` GUI, which is where messages are collected and shown in a table that consists of individual columns for the timestamp, the message itself, the severity level, and the node that produced the message, along with other information.

This interface allows us to pause, save, and load previously saved logging messages. We can also clear the list of messages and filter them by using highlighting filters.

Moreover, we can adjust the columns automatically by pressing the **Resize columns** button and see all the information regarding them, including the line of code that was generated, which is shown in the following screenshot:

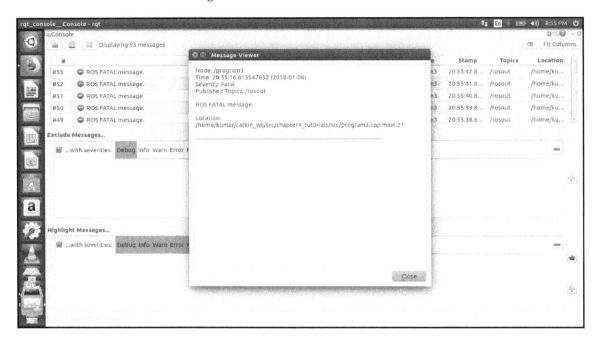

GUI tools interface

In order to set the severity of the loggers, we can use the `rqt_logger_level` GUI tool:

```
$ rosrun rqt_logger_level rqt_logger_level
```

The output is shown in the following screenshot:

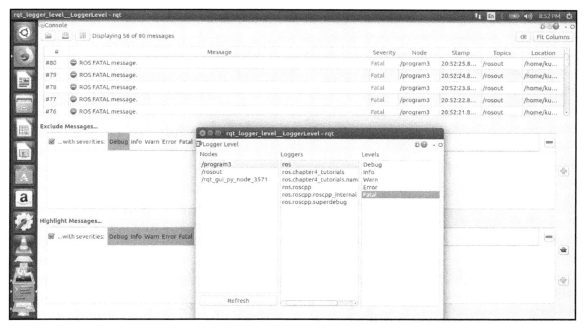

GUI tool rqt_logger_level

As shown in the preceding screenshot, we can select the node, followed by the named logger, and finally its severity. After modification, the new messages received with a severity below the desired level will not appear in the `rqt_console` GUI.

We can observe that each node has several internal loggers by default, most of which are related to the ROS communication API. Usually, it is not recommended to reduce their severity.

Inspecting and diagnosing the ROS system

While the system is running, there are several nodes publishing messages and there are nodes providing actions or services. In the case of a very large system, it becomes important to have tools that can provide information about the running system. However, the ROS provides a set of basic and very powerful tools for system inspection, from the CLI to GUI applications.

Getting ready

In Chapter 2, *ROS Architecture and Concepts – I*, we discussed listing nodes, topics, services, and parameters. The CLI commands `rosnode list`, `rostopic list`, `rosservice list`, and `rosparam list` provide the most basic level of introspection. Any of these commands can also be combined with regular bash commands to look for the desired nodes, topics, services, or parameters, such as `grep`:

```
$ rostopic list | grep speed
```

How to do it...

ROS provides several powerful GUI tools for system introspection. First of all, we will discuss `rqt_top`, which shows nodes and resources they are using, which is very similar to the `top` command in the Linux system. The following screenshot shows the `rqt_top` command for the `program4` and `program5` nodes running in the system:

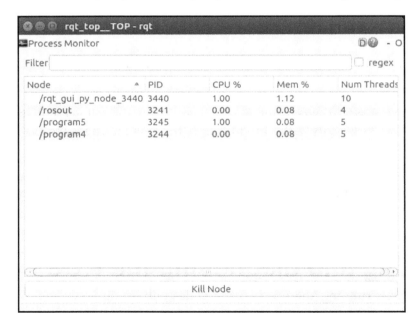

rqt_top

Additionally, `rqt_topic` provides information about the topics, which includes publishers, subscribers, the publishing rate, and the messages being published. We can view the message fields and select the topic for introspection by using the following code:

```
$ rosrun rqt_topic rqt_topic
```

The following screenshot shows the GUI and the workings of the `rqt_topic` tool:

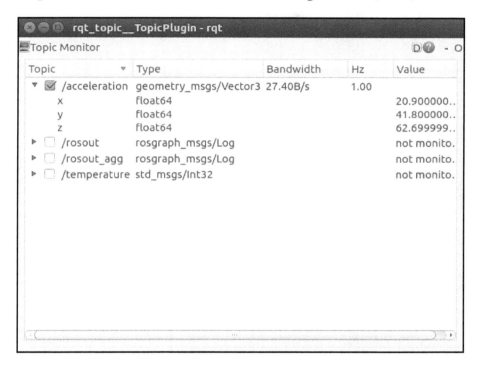

rqt_topic

Moving forward, we can see `rqt_publisher`, which helps us in managing multiple instances of `rostopic` pub commands in a single GUI interface, as shown in the following screenshot:

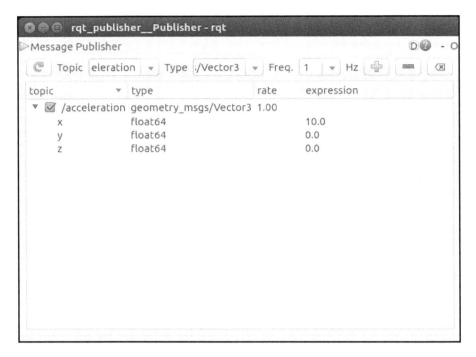

rqt_publisher

Similarly, `rqt_service_caller` allows us to manage multiple instances of `rosservice` call commands in a single interface, which is shown in the following screenshot:

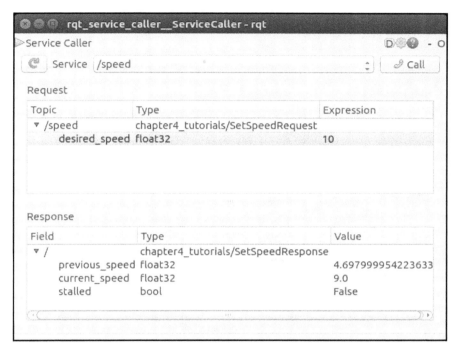

rqt_service_call

The current state of an ROS session can be represented as a directed graph where graph nodes correspond to the running node and edges are the publisher-subscriber connections. The ROS tool `rqt_graph` can be used in dynamically drawing this graph:

```
$ rosrun rqt_graph rqt_graph
```

The following screenshot shows the directed graph for our example, where `program4` and `program5` are running in the current session:

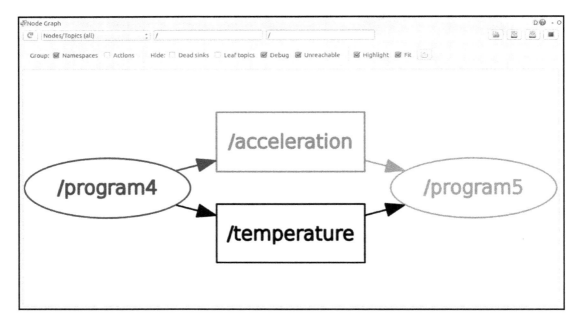

rqt_graph

In the preceding screenshot, we have the `program4` and `program5` nodes connected by the topic's temperature and acceleration. The `rqt_graph` GUI provides several options for customizing the view.

Finally, we will learn how to enable statistics to observe the message rate and bandwidth. We have to set the statistics parameter before running `rqt_graph` in order to make this information available:

```
$ rosparam set enable_statistics true
```

Hopefully, we will get the information we need by using `rqt_graph`, which is shown in the following screenshot:

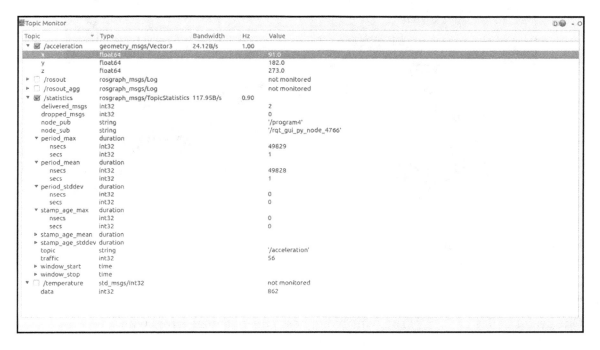

Statistics with rqt_graph

The ROS provides the `roswtf` tool, which statically analyzes a given package and detects potential problems. For `chapter4_tutorials`, we have the following output:

```
$ roscd chapter4_tutorials $ roswtf
```

The output is shown in the following screenshot:

```
kumar@kumar-Inspiron-5437:~/catkin_ws/src/chapter4_tutorials$ roswtf
Loaded plugin tf.tfwtf
Package: chapter4_tutorials
================================================================================
Static checks summary:

No errors or warnings
================================================================================
Beginning tests of your ROS graph. These may take awhile...
analyzing graph...
... done analyzing graph
running graph rules...
... done running graph rules

Online checks summary:

No errors or warnings
kumar@kumar-Inspiron-5437:~/catkin_ws/src/chapter4_tutorials$
```

roswtf analysis report

In the preceding screenshot, we can observe that `roswtf` does not detect any errors, but even then some of them are innocuous. However, the purpose of `roswtf` is to signal potential problems; it is our responsibility to verify whether they are important to consider or not.

One useful tool is `catkin_lint`, which diagnoses the potential errors with `catkin`, generally, in the `CMakeLists.txt` and `package.xml` files:

```
$ catkin_lint -W2 --pkg
```

The `catkin_lint` might need to be installed separately, which is usually contained in the `python-catkin-lint` package.

In the following screenshot, we can see many warnings for `CMakeLists.txt` in the `chapter4_tutorials` package; it is our responsibility to verify whether they are important to consider or not:

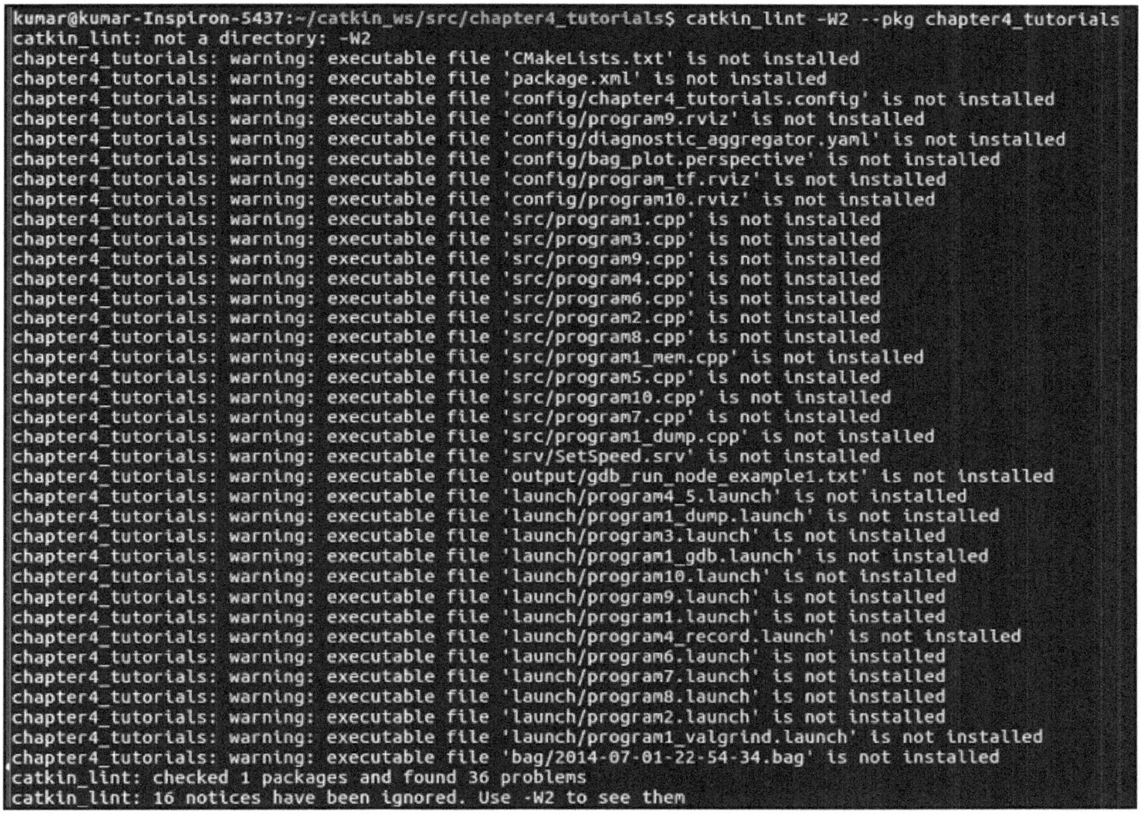

```
kumar@kumar-Inspiron-5437:~/catkin_ws/src/chapter4_tutorials$ catkin_lint -W2 --pkg chapter4_tutorials
catkin_lint: not a directory: -W2
chapter4_tutorials: warning: executable file 'CMakeLists.txt' is not installed
chapter4_tutorials: warning: executable file 'package.xml' is not installed
chapter4_tutorials: warning: executable file 'config/chapter4_tutorials.config' is not installed
chapter4_tutorials: warning: executable file 'config/program9.rviz' is not installed
chapter4_tutorials: warning: executable file 'config/diagnostic_aggregator.yaml' is not installed
chapter4_tutorials: warning: executable file 'config/bag_plot.perspective' is not installed
chapter4_tutorials: warning: executable file 'config/program_tf.rviz' is not installed
chapter4_tutorials: warning: executable file 'config/program10.rviz' is not installed
chapter4_tutorials: warning: executable file 'src/program1.cpp' is not installed
chapter4_tutorials: warning: executable file 'src/program3.cpp' is not installed
chapter4_tutorials: warning: executable file 'src/program9.cpp' is not installed
chapter4_tutorials: warning: executable file 'src/program4.cpp' is not installed
chapter4_tutorials: warning: executable file 'src/program6.cpp' is not installed
chapter4_tutorials: warning: executable file 'src/program2.cpp' is not installed
chapter4_tutorials: warning: executable file 'src/program8.cpp' is not installed
chapter4_tutorials: warning: executable file 'src/program1_mem.cpp' is not installed
chapter4_tutorials: warning: executable file 'src/program5.cpp' is not installed
chapter4_tutorials: warning: executable file 'src/program10.cpp' is not installed
chapter4_tutorials: warning: executable file 'src/program7.cpp' is not installed
chapter4_tutorials: warning: executable file 'src/program1_dump.cpp' is not installed
chapter4_tutorials: warning: executable file 'srv/SetSpeed.srv' is not installed
chapter4_tutorials: warning: executable file 'output/gdb_run_node_example1.txt' is not installed
chapter4_tutorials: warning: executable file 'launch/program4_5.launch' is not installed
chapter4_tutorials: warning: executable file 'launch/program1_dump.launch' is not installed
chapter4_tutorials: warning: executable file 'launch/program3.launch' is not installed
chapter4_tutorials: warning: executable file 'launch/program1_gdb.launch' is not installed
chapter4_tutorials: warning: executable file 'launch/program10.launch' is not installed
chapter4_tutorials: warning: executable file 'launch/program9.launch' is not installed
chapter4_tutorials: warning: executable file 'launch/program1.launch' is not installed
chapter4_tutorials: warning: executable file 'launch/program4_record.launch' is not installed
chapter4_tutorials: warning: executable file 'launch/program6.launch' is not installed
chapter4_tutorials: warning: executable file 'launch/program7.launch' is not installed
chapter4_tutorials: warning: executable file 'launch/program8.launch' is not installed
chapter4_tutorials: warning: executable file 'launch/program2.launch' is not installed
chapter4_tutorials: warning: executable file 'launch/program1_valgrind.launch' is not installed
chapter4_tutorials: warning: executable file 'bag/2014-07-01-22-54-34.bag' is not installed
catkin_lint: checked 1 packages and found 36 problems
catkin_lint: 16 notices have been ignored. Use -W2 to see them
```

catkinlint ouputs

Visualizing and plotting scalar data

ROS already has generic tools to plot scalar data. These tools can also be used for plotting non-scalar data, but each scalar field needs to be plotted separately in order to show less information. Alternately, ROS has powerful visualization tools for non-scalar or vector data, which we will discuss in the next section.

Getting ready

In this section, we will learn that scalar data can be plotted as a time series, where time is provided by the timestamps of the messages. We can use the *y* axis to plot scalar data versus timestamps along the *x* axis. ROS has a GUI tool called rqt_plot, which will do this for us.

To see rqt_plot in action, we will use the program4 node which publishes scalar and vector data in two different topics, /temperature and /acceleration, respectively (refer to program4.cpp at GitHub for more information). However, the values in the message are synthetically generated and so they do not have physical meaning but are useful in our plotting demonstrations and learning.

How to do it...

We will start our discussion by running the program4 and program5 nodes using the following command:

```
$ roslaunch chapter4_tutorials program4_5.launch
```

1. To plot a message using rqt_plot

    We must have information about the type and structure of the messages; we can use rosmg show <msg type> to get this.

2. Run the following command to get the plot for the message in the /temperature topic, which is of the Int32 type of scalar data:

    ```
    $ rosrun rqt_plot rqt_plot /temperature/data
    ```

3. Select the message topic by the menu from the rqt_plot GUI after starting rqt_plot:

    ```
    $ rosrun rqt_plot rqt_plot
    ```

While the node is running and when `rqt_plot` is started, we will see a plot that changes over time with the incoming messages from the `/temperature` topic, as shown in the following screenshot:

rqt_plot scalar data

Similarly, for the `/acceleration` topic, we have a Vector3 message (`rostopic type /acceleration`), which has three fields that we can visualize in a single plot as follows:

```
$ rosrun rqt_plot rqt_plot /acceleration/x:y:z
```

Alternatively, we can also select the `/acceleration` messages fields from the GUI menu, after launching the `rqt_plot` tool. Hopefully, we will see a plot similar to the one in the following screenshot:

rqt_plot non-scalar data

There's more...

Moreover, we can also plot each field in a separate window that is not supported by `rqt_plot` directly. Nevertheless, we can use the `rqt_gui` tool and plot each field in parallel in the separate window, which can be arranged manually after starting the `rqt_gui` tool, as follows:

```
$ rosrun rqt_gui rqt_gui
```

We will get the following output:

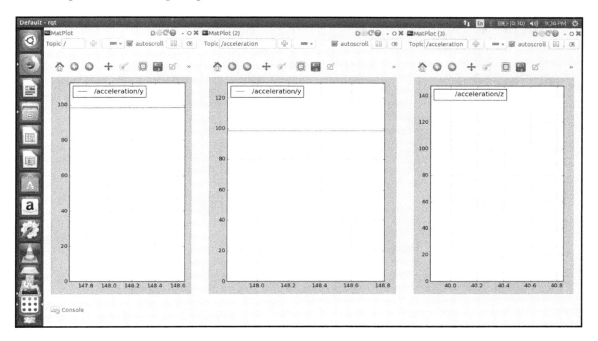

rqt_gui plotting non-scalar data

The `rqt_plot` supports plotting three GUI frontends, as shown in the following screenshot. We can access and select them from the configuration button.

However, we will select the **MatPlot** frontend since it supports more time series simultaneously:

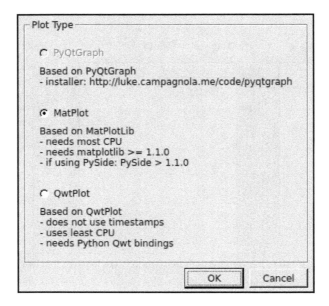

rqt_plot GUI frontends

Visualizing non-scalar data – 2D/3D images

There are several robotic applications that deal with 2D and 3D data, usually in the form of images or point cloud. ROS provides interfaces that are built upon the OpenCV library to solve computer vision problems for complex robotics applications.

Getting ready

There are several extremely useful tools available in ROS to visualize these types of data, such as `image_view` and `rqt_image_view` for 2D data. Similarly, in ROS, we have `rviz` and `rqt_rviz`, which integrates an OpenGL interface with a 3D world that represents sensor data in a world representation.

How it works...

In this section, we will discuss `program8` (refer
to `chapter4_tutorials/src/program8.cpp` on GitHub), which implements a basic
camera capture program using OpenCV and ROS binding to convert `cv::Mat` images into
ROS image messages. This node publishes the image messages from camera frames in the
`/camera` topic and enables us to display images coming from a camera on-the-fly. We will
discuss more about source code and implementation in the upcoming chapter, or interested
readers could refer to the online resources (`https://docs.opencv.org/3.0-beta/modules/
videoio/doc/reading_and_writing_video.html`).

However, we can run the node using the `launch` file:

```
$ roslaunch chapter4_tutorials program8.launch
```

We can't use `rostopic echo /camera`, since the amount of information in the plain text
would be huge and cannot be read by humans. Hence, ROS provides the `image_view`
node, which shows the images in the specific topic in a window:

```
$ rosrun image_view image_view image:=/camera
```

By pressing the right button of the mouse in the window, we can save the current frame to
the disk, usually in `~/.ros`.

Moreover, ROS Kinetic provides the `rqt_image_view` tool, which supports viewing
multiple images in a single window but does not support saving images by right-clicking
on them:

```
$ rosrun rqt_image_view rqt_image_view
```

Here, we can select the image topic manually from the GUI in the drop-down menu, as shown in the following screenshot:

ROS rqt_image_view

ROS provides several interfaces that are built on top of the OpenCV for solving computer vision problems. We can refer to the online resources for more details, but this is out of the scope of this book.

We discussed RViz and plugin development in the previous chapter. Nevertheless, we will learn how to visualize 3D data in ROS using `rqt_rviz`, which is a powerful 3D visualization tool. We will start `rqt_rviz` while `roscore` is running to see the graphical interface of the following screenshot, which is a simple layout:

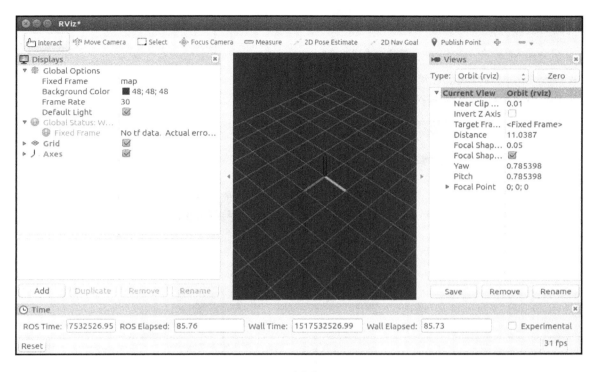

rqt_rviz display

Here, on the top left-hand side, we have the **Displays** panel, which has a tree list of several elements in the world. However, in this case, we can observe a set of elements that have already loaded into the layout, which is saved in the `config/program9.rviz` file. We have to load the elements into the **File | Open Config** menu.

Additionally, we have the **Add** button below the **Displays** panel, which allows us to add more elements by topics and their type. In the case of `program9`, as an example, we would add `Marker` and `PointCloud2`. We also have a menu bar for configuring **2D Nav Goal**, **2DPose Estimate**, **Interact**, **Move Camera**, **Measure**, and so on. Similarly, the **Views** menu provides different view types, such as **Orbit**, **TopDownOrtho**, and so on, with some other information regarding viewing the parameter.

At this moment in time, we will start the `program9` node using the following command:

```
$ roslaunch chapter4_tutorials program9.launch
```

In `rqt_rviz`, we will set the fixed frame to `frame_marker`, which will look like a red cubic marker is moving, as shown in the following screenshot:

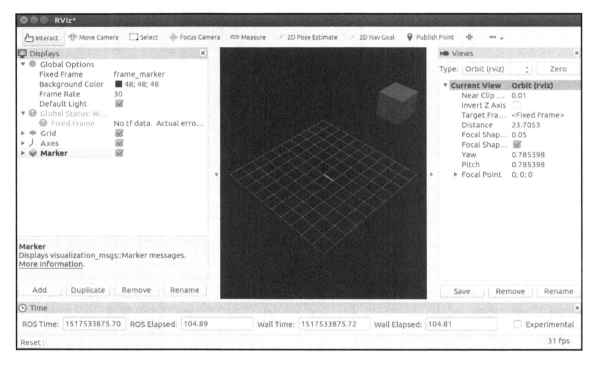

rqt_rviz marker display

Similarly, if we set the fixed frame to `frame_robot`, we will see a rainbow of point cloud panes, as shown in the following screenshot:

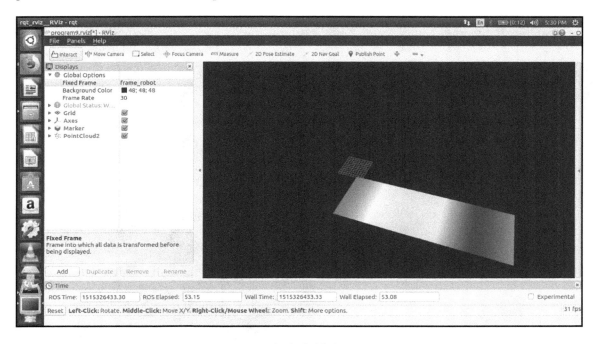

rqt_rviz point cloud display

`rqt_rviz` also provides the interactive marker:

```
$ roslaunch chapter4_tutorials program10.launch
```

We have the `program10` node, which shows a simple interactive marker, as shown in the following screenshot:

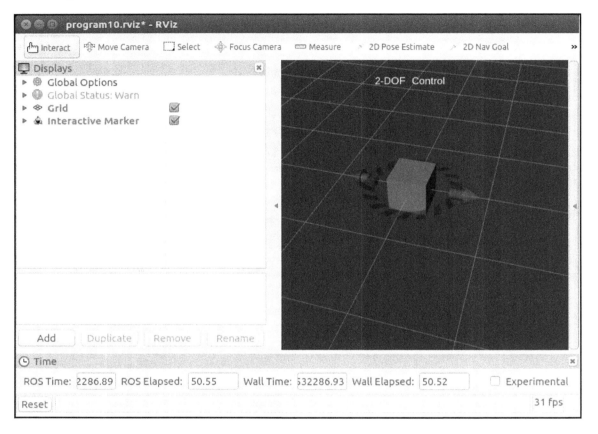

rqt_rviz interactive marker

Here, we will notice a marker that can move in the interactive mode of `rqt_rviz`. It is an example program that shows how its pose can be used to modify the pose of another element in the system, such as the joint or position of a robot.

In the previous chapter, we discussed the ROS **Transform Frame** (**TF**), where we were able to visualize the frame of the lead and follower turtle in Rviz, as shown in the following screenshot:

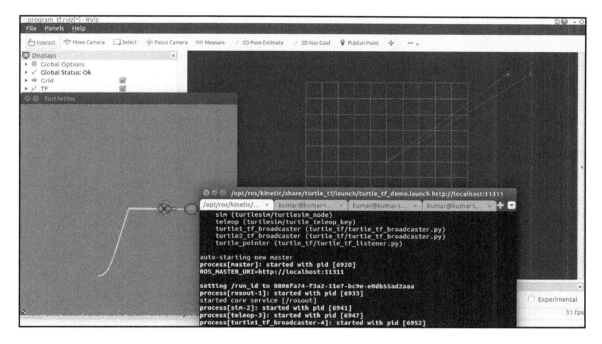

Transform Frame (TF)

`rqt_rviz` is an unusually powerful visualization tool, and it would not be possible to discuss all of its features and properties in the limited pages of this book. However, as a starter, we will try and provide an overview of the generic features. The interested and advanced reader should refer to the online resources for more information.

Recording and playing back ROS topics

In general, when we are developing complex robotics systems, not all resources are available all the time because of the cost or time required for preparing and performing experiments. Therefore, it is good practice to record the data of the experiment session for later analysis.

However, the process of saving experiment data and reproducing the experiment offline for development and analysis is not trivial. The ROS development environment provides powerful tools, also known as `rosbag`, to address such problems. This enables us to reproduce the experiment offline, with its real conditions on the robot such as the latency of transmitting messages being efficient, with a high bandwidth, and in an adequate and organized manner.

In this section, we will discuss how to use these ROS tools to save and playback data that is stored in bag files, a binary format designed for ROS developers. We will also learn how to manage these files, inspect their content, compress them, and split or merge them.

Getting ready

A bag file is a container of messages on ROS topics published during the execution of the system, and this allows us to play them back virtually as a real system, even with time delays, since all messages are recorded with a timestamp and a header.

The ROS bag file stores data in a particular binary format for extremely fast recording throughput. Additionally, the size of the bag file is very important. Nonetheless, there is an option to compress the file on the fly with the `bz2` algorithm using the `-j` parameter. Since each and every message is recorded along with the topic and publisher, this enables us to selectively specify which topics to record or record them all using the `-a` parameter. Similarly, we can also select a particular subset of the topics contained in the bag file by specifying the name of the topics that need to be published.

How it works...

First of all, we will learn how to record some data as an example—we will use our `program4` node, which was discussed in the previous section *Inspecting and diagnosing the ROS system* . We will first launch the node as follows:

```
$ roslaunch chapter4_tutorials program4.launch
```

At the moment, we have two options: first, to record all the topics, and second, to record selective topics as follows, respectively:

```
$ rosbag record -a $ rosbag record /temperature
```

Once the experiment is completed, we can terminate the recording by hitting *Ctrl + C* in the running terminal.

We can explore more options with the `rosbag` help record, which includes the bag file size, the duration of the recording, options to split the files into several smaller files of a given size, compressing the file on the fly, and so on. During recording, if messages are dropped, we could increase the buffer (–b parameter) size for the recorder from 256 MB (default size) to a couple of GB if the data rate is very high.

It is also good practice to include the call to the `rosbag` record into a `launch` file, which is useful in setting up a recorder for certain topics, as follows for the example node (refer to `program4_record.launch` at GitHub):

```
<launch> <!-- Program 4 --> <node pkg=""chapter4_tutorials""
type=""program4"" name=""program4"" output=""screen""/> <!-- Bag record -->
<node pkg=""rosbag"" type=""record"" name=""bag_record""
args=""/temperature /acceleration""/> </launch>
```

The bag file is created by default in ~/.ros while running from the `launch` file. Otherwise, we can configure this by specifying the name of the file with –o (prefix) or –O (full name).

Now, we have a `rosbag` file recorded, which can be used for playing back all messages from their respective topics. We will run the `rosbag` file as follows:

```
$ rosbag play 2018-01-07-22-54-35.bag
```

We can see that `rosbag` will produce the following output:

```
kumar@kumar-Inspiron-5437:~/.ros$ rosbag play -r 100 2018-01-07-10-35-00.bag
[ INFO] [1515301638.127694416]: Opening 2018-01-07-10-35-00.bag

Waiting 0.2 seconds after advertising topics... done.

Hit space to toggle paused, or 's' to step.
 [RUNNING]  Bag Time: 1515301515.839165    Duration: 14.021838 / 16.000160
Done.
```

Running rosbag

We can pause this by hitting the *space bar* or move step by step by hitting *S*. Generally, you should press *Ctrl* + *C* to finish the execution. However, it will close automatically once it reaches the end of the file. There is an option to loop (`-l` parameter), which might be useful in some cases.

Moreover, we can see the topics with the `rostopic` list as follows:

```
kumar@kumar-Inspiron-5437:~/catkin_ws$ rostopic list
/acceleration
/clock
/rosout
/rosout_agg
/temperature
kumar@kumar-Inspiron-5437:~/catkin_ws$ ▮
```

rosbag topic list

Here, we can see the `/clock` topic, which simulates the system clock to produce a faster or slower playback, and could be configured using the `-r` option. Furthermore, we can also specify a subset of the topics to be published by using the `--topics` option.

There are two major methods that can be used to inspect the contents of a bag file. The first one is from the command-line interface: `rosbag info <bag_file>`. We will see something similar to what is shown in the following screenshot:

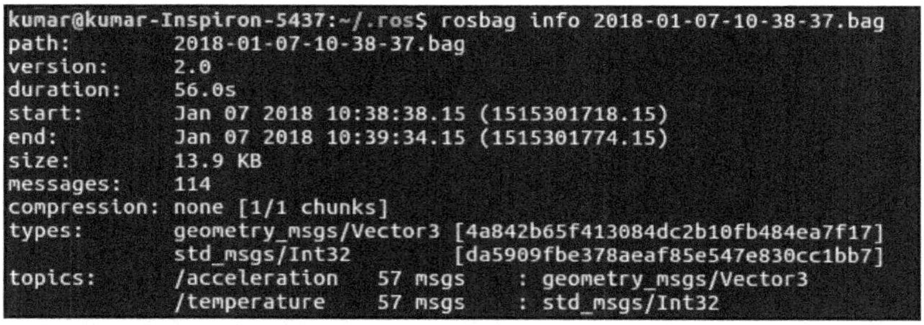

```
kumar@kumar-Inspiron-5437:~/.ros$ rosbag info 2018-01-07-10-38-37.bag
path:         2018-01-07-10-38-37.bag
version:      2.0
duration:     56.0s
start:        Jan 07 2018 10:38:38.15 (1515301718.15)
end:          Jan 07 2018 10:39:34.15 (1515301774.15)
size:         13.9 KB
messages:     114
compression:  none [1/1 chunks]
types:        geometry_msgs/Vector3 [4a842b65f413084dc2b10fb484ea7f17]
              std_msgs/Int32        [da5909fbe378aeaf85e547e830cc1bb7]
topics:       /acceleration   57 msgs    : geometry_msgs/Vector3
              /temperature    57 msgs    : std_msgs/Int32
```

rosbag information

It shows information about the bag file itself, such as the creation date, duration, size, the number of messages with the associated topics, and the compression, if any.

The second method to inspect a bag file is extremely powerful. This method is to use a GUI tool called `rqt_bag`, which allows us to play back the files, view the images, plot scalar data, and also view the raw structure of the messages. We can select the name of the bag file from GUI for our experiment. Hopefully, it will produce the following output:

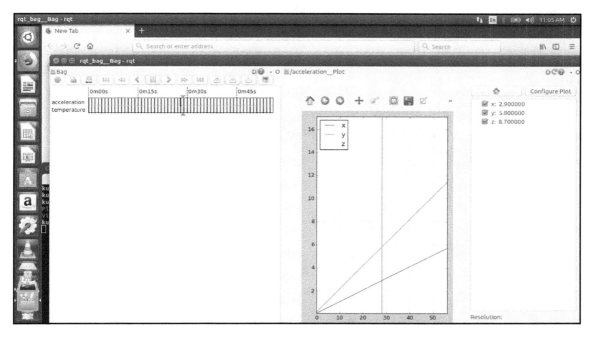

rqt_bag

Here, we have a timeline for all the topics where each message appears with a mark. The bag file also contains images that can be enabled in thumbnails so that you can view them in the timeline.

As an advanced user, we can configure the `rqt_gui`, `rqt_bag`, and `rqt_plot` plugins in the same window where the layout can be imported from the perspective defined in `bag_plot.perspective` (refer to `/chapter4_tutorials/config/bag_plot.perspective` at GitHub for more information).

There's more...

`rosbag compress` is a command-line tool that is used for compressing bag files. Currently, there are two supported formats—BZ2 and LZ4. However, BZ2 is the default. We can get its uses and help by using the following command (refer to `http://wiki.ros.org/rosbag/Commandline#compress` for more information):

```
$ rosbag compress -h
```

And compress the given bag files using BZ2 as follows:

```
$ rosbag compress *.bag
```

Similarly, `rosbag decompress` is a command-line tool for decompressing bag files. Moreover, it automatically determines which compression format a bag uses. We can get the uses and help by using the `-h` option, as follows:

```
$ rosbag decompress -h
```

And decompress the given bag files using the following command:

```
$ rosbag decompress *.bag
```

We can also record all active topics in `rosbag` to analyze the system. However, we can exclude some of the topics such as point cloud or image topics from recording by using the `-x` option:

```
$ rosbag record -a -x ""/usb_cam/(.*)|/usb_cam_repub/theora/(.*)""
```

Moreover, `rosbag` records the file in chunks of a given size:

```
$ rosbag record -a --size=500 -x
""/usb_cam/(.*)|/usb_cam_repub/theora/(.*)""
```

Similarly, **rosbag** records the file in chunks of a given duration:

```
$ record -a --duration 2m --duration 10m -x
""/usb_cam/(.*)|/usb_cam_repub/theora/(.*)""
```

5
Accessing Sensors and Actuators through ROS

In this chapter, we will discuss the following recipes:

- Understanding the Arduino-ROS interface
- Interfacing 9DoF Razor IMU-Arduino-ROS
- Using a GPS system – Ublox
- Interfacing servomotors – Dynamixel
- Using a laser rangefinder – Hokuyo
- Working with the Kinect sensor to view objects in 3D
- Using a joystick or a gamepad in ROS

Introduction

In the previous chapters, we discussed the visualization and debugging tools that are used in ROS during development. In this chapter, we will learn about interfacing hardware components, such as sensors and actuators, to the ROS framework. First, we will look at interfacing sensors using I/O boards, such as Arduino, and interfacing the 9DoF Razor IMU. Similarly, we will also discuss interfacing actuators, Dynamixel, the laser rangefinder – Hokuyo, a GPS system – Ublox, and the RGB-D – Kinect Sensor for viewing objects in 3D.

Since it is impossible to explain all the types of available sensors and actuators in one chapter, we have selected some of the most commonly used sensors and actuators found in robotics laboratories to start learning. These sensors and actuators can be organized into different categories: rangefinders, perception, pose estimation devices, and so on.

Understanding the Arduino-ROS interface

We are going to look at the Arduino-ROS interface first. Arduino is an open-source electronics platform based on easy-to-use hardware and software (`https://www.arduino.cc/`). Arduino perceives the environment by receiving inputs from many sensors and affects its surroundings by controlling lights, motors, and other actuators. Arduino can be used for the quick prototyping of robots. The main applications of Arduino in robotics are interfacing sensors and actuators for the computer system using the UART interface. Most of the Arduino boards are powered by Atmel microcontrollers, which are available from 8-bit to 32-bit, with clock speeds from 8 MHz to 84 MHz.

There are a variety of Arduino boards available on the market. However, selecting one of them is dependent on the nature of the robotic application you're developing. Some of them include Arduino UNO, Arduino Mega, and Arduino DUE, which are categorized as beginner, intermediate, and high-end, respectively.

Getting ready

The Arduino-ROS interface is a standard method of communication between the Arduino boards and the PC, which uses the UART protocol. When both the devices communicate with each other, there should be some program running on both sides that can interpret the serial messages for each of these devices.

In this section, we will discuss the interface that's exclusive to Arduino. However, we can implement our own logic to receive and transmit the data from the board to PC and vice versa. The interfacing code will be different in each I/O board because there are no standard libraries to perform this communication.

Arduino IDE is also used for programming the Arduino board, and this can be downloaded from `https://www.arduino.cc/en/Main/Software`.

The `rosserial` package is a protocol for wrapping standard ROS serialized messages and multiplexing multiple topics and services over a character device such as a serial port or network socket. In other words, the rosserial protocol can convert the standard ROS messages and services data types to embedded device equivalent data types, such as a UART data packet.

The ROS client libraries, such as `roscpp`, `rospy`, and `roslisp`, enable us to develop ROS nodes that you can up and run on various systems. These client libraries are ports of the general ANSI C++ `rosserial_client` library. Currently, these packages include:

- `rosserial_arduino`: Arduino, especially UNO and Leonardo, but also Arduino MEGA and DUE, Teensy 3.x, and LC, Spark, STM32F1, and ESP8266-based boards
- `rosserial_embeddedlinux`: Embedded Linux platform
- `rosserial_windows`: Communicating with Windows applications
- `rosserial_mbed`: Mbed platforms
- `rosserial_tivac`: Tiva C Launchpad boards from TI's Launchpad ecosystem—TM4C123GXL and TM4C1294XL
- `rosserial_stm32`: STM32 MCUs
- `ros-teensy`: Teensy platforms

We require some other packages to decode the serial message from the `rosserial_client` libraries and convert them into topic and service messages format for the ROS framework that's running on the computer system. The following packages help in decoding the serial data:

- `rosserial_python`: This is the recommended computer system-side package for handling serial data from a device. This node is completely written in Python.
- `rosserial_server`: This is a C++ implementation of `rosserial` in the computer system side. Although the inbuilt functionalities are less compared to `rosserial_python`, it can be used for high-performance applications.

How to do it...

1. To run the ROS nodes from Arduino, let us set up the `rosserial` packages and the `rosserial_arduino` client in the Arduino IDE.
2. Install the `rosserial` packages on Ubuntu 16.04 using the following commands:

```
$ sudo apt-get install ros-kinetic-rosserial-arduino ros-kinetic-
rosserial-embeddedlinux ros-kinetic-rosserial-windows ros-kinetic-
rosserial-server ros-kinetic-rosserial-python
```

3. Build the packages from the source after cloning the `rosserial` repository in the catkin workspace:

```
$ cd ~/catkin_ws/src/
$ git clone https://github.com/ros-drivers/rosserial.git
$ cd ~/catkin_ws/
$ catkin_make
```

4. Install the Arduino IDE by following the steps described at http://arduino.cc/en/Main/Software.

5. Download the Linux 64-bit version and copy the Arduino IDE folder to the Ubuntu desktop.

   Arduino requires Java runtime support to run it. If you do not have Java installed, we can install it using the following command:

   ```
   $ sudo apt-get install java-common
   ```

6. After installation, we can start the Arduino IDE from the Ubuntu application launcher. The following screenshot shows the Arduino IDE window:

Arduino IDE window

How it works...

After setting up the Arduino, we will go to **File** | **Preferences** for configuring the sketchbook folder of Arduino (for example, /home/<user>/Arduino).

You may have noticed a folder called libraries inside the Arduino folder. There, we have to generate ros_lib using a script called make_libraries.py, which is present inside the rosserial_arduino package. The following commands show how to generate ros_lib for Arduino:

 ros_lib is the rosserial_client for Arduino, which provides the ROS client APIs inside an Arduino IDE environment.

```
$ cd ~/Arduino/libraries/
$ rosrun rosserial_arduino make_libraries.py
```

The make_libraries.py script will generate a wrapper of the ROS topic and service messages, which are optimized for Arduino data types, and generate a folder called ros_lib inside the libraries folder. If we restart the Arduino IDE, we will see that ros_lib appears in the **File** | **Examples** tab.

We can take an example and make sure that it is building properly to ensure that the ros_lib APIs are working fine. Next, we will discuss the basic program structure of the ROS Arduino node and the necessary APIs:

```
#include <ros.h>
ros::NodeHandle node;
void setup() {
  node.initNode();
}
void loop() {
  node.spinOnce();
}
```

The preceding code snippet shows the program structure of the Arduino node, where Nodehandle should be declared before the setup() function, which will give a global scope to the NodeHandle instance called node. The initialization of this node should be done inside the setup() function. Note that the Arduino setup() function will only execute once when the device starts. It is important to know that we can only create one node for a serial device.

Moreover, inside the `loop()` function, we have to use `spinOnce()` to invoke the ROS callback:

```
ros::Subscriber<std_msgs::String> listner("listener", callback);
ros::Publisher talker("talker", &str_msg);
```

After defining the publisher and the subscriber, we have to initiate this inside the `setup()` function as follows:

```
node.advertise(chatter);
node.subscribe(sub);
```

We can find more reference code from `ros_lib` example programs.

Now, we will discuss our first example using the Arduino and ROS interfaces as listener and `talker` interfaces. Users can send a `String` message to the `talker` topic and Arduino will publish the same message in a `listener` topic.

The following source code implements a ROS node for Arduino:

```
#include <ros.h>
#include <std_msgs/String.h>
//Creating Nodehandle
ros::NodeHandle  node;
std_msgs::String str_msg;
//Defining Publisher and callback
ros::Publisher talker("talker", &str_msg);
void callback ( const std_msgs::String& msg){
 str_msg.data = msg.data;
 talker.publish( &str_msg );
}
//Defining Subscriber
ros::Subscriber<std_msgs::String> listener("listener", callback);
void setup()
{
  //Initializing node
  node.initNode();
  //Configure Subscriber and Publisher
  node.advertise(talker);
  node.subscribe(listener);
}
void loop()
{
  node.spinOnce();
  delay(2);
}
```

We can compile the preceding code and upload it to the Arduino board. After uploading the code, we will select the desired Arduino board that we are using for this example and the device serial port of the Arduino IDE. More details can be found at `https://www.arduino.cc/en/Guide/HomePage`. (Go to **Tools** | **Boards** to select the board and `Tools` | `Port` to select the device port name of the board.)

After compiling and uploading the code to the Arduino board, we will have to start the ROS bridge nodes on the Linux system that connects Arduino and the Linux system.

We will create an `arduino.launch` file inside the `rosserial_python` packages with the following contents:

```
<launch>
<node pkg = "rosserial_python" type="serial_node.py" name="serial_node"
output="screen">
        <param name="~port" value="/dev/ttyACM0" />
        <param name="~baud" value="57600" />
</node>
</launch>
```

Before launching and getting the configuration parameter for `arduino.launch`, we have to ensure that Arduino is already connected to the Linux system.

We are using the `rosserial_python` node here as the ROS bridging node. We have to mention the device name port and baud-rate as a parameter. For the device name port, we have to search for the port name listing from the contents of the `/dev` directory. We are using the `/dev/ttyACM0` port here. Note that root permissions are required to use this port. However, we can change the permissions using the following command in order to read and write data on the desired port:

```
$ sudo chmod 666 /dev/ttyACM0
```

Where `666` allows read and write permissions to all. The default baud-rate of this communication is 57,600. However, we can change the baud-rate according to our application. The usage of `serial_node.py` inside the `rosserial_python` package is given at `http://wiki.ros.org/rosserial_python`.

Now, we can launch `arduino.launch` and start the communication between the Linux system and the Arduino board:

```
$ roslaunch rosserial_python arduino.launch
```

When `serial_node.py` starts running from the Linux system, it will send some serial data packets called query packets to get the number of topics, the topic names, and the types of topics that are received from the Arduino node.

If there is no response for the query packet, it will send it again. However, the synchronization in communication is based on ROS time.

The buffer size allocated for publishing and subscribing is 512 bytes by default. However, the buffer allocation is dependent on the amount of RAM available on the particular microcontroller that we are working with. We can override these settings by changing the `BUFFER_SIZE` macro inside `ros.h`.

After running `serial_node.py`, we will get a list of topics by using the following command:

```
$ rostopic list
```

 You may have noticed that topics such as `/listener` and `/talker` are being generated. We can simply publish a message to the `/listener` topic by using the following command:

```
$ rostopic pub -r 5 listener std_msgs/String "Hello Arduino"
```

We can echo the `/talker` topic, and we will get the same message as we have published:

```
$ rostopic echo /talker
```

Great! This shows that we have established the communication channel between the Linux system and the Arduino board. In the next section, we will look into some more applications that use the Arduino-ROS interface.

Interfacing 9DoF Razor IMU-Arduino-ROS

In this section, we will learn how to use a low-cost sensor with the 9 **Degrees of Freedom (DoF)** Razor IMU M0 with the ROS framework. The SparkFun 9DoF Razor IMU M0 combines a SAMD21 microprocessor with an MPU-9250 9DoF (nine degrees of freedom) sensor to create a tiny, re-programmable, multi-purpose **inertial measurement unit (IMU)**.

The 9DoF Razor IMU M0 incorporates three sensor: an accelerometer, a gyroscope, and a magnetometer, which gives it the ability to sense linear acceleration, angular rotation velocity, and magnetic field vectors.

The onboard microprocessor, Atmel's SAMD21G18A, is an Arduino-compatible, 32-bit ARM Cortex-M0+ microcontroller, which is also featured on the Arduino Zero and SAMD21 Mini Breakout boards. We can update the firmware or develop our own code easily using Arduino IDE. Alternatively, we can also use our own Arduino to control the sensor stick using I2C.

Getting ready

The 9DoF Razor IMU M0 is designed around the SAMD21—the same processor on the Arduino Zero, which means adding Arduino support for the board is just a few clicks away.

As we mentioned earlier, the sensor stick can be controlled using the I2C protocol, and therefore we will use Arduino to control it. The following diagram shows the connection diagram between the Arduino UNO and 9DoF Razor IMU boards. We can refer to the online material from Sparkfun for more details on setup and connections (`https://learn.sparkfun.com/tutorials/9dof-razor-imu-m0-hookup-guide`):

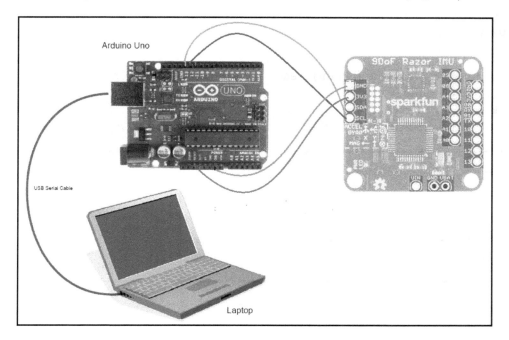

Arduino UNO and 9DoF Razor IMU M0 board connection

How to do it...

1. Install the Razor IMU ROS library to establish the communication interface.

2. Install visual Python using the following command:

```
$ sudo apt-get install python-visual
```

3. Download and build the `razor_imu_9dof` packages in our workspace from the repository:

```
$ cd ~/catkin_ws/src
$ git clone https://github.com/KristofRobot/razor_imu_9dof.git
$ cd ..
$ catkin_make
```

How it works...

Here, we are ready with the Linux system ROS-9DoF Razor IMU interface software packages. We also need the Arduino interface software, which can control the 9DoF Razor IMU using the I2C protocol and also enable the Arduino-ROS interface to communicate with the Linux system.

We will launch the Arduino IDE installed in the previous section and open the `Razor_AHRS.ino` file from the `razor_imu_9dof` package at `~/catkin_ws/src/razor_imu_9dof/Razor_AHRS/Razor_AHRS.ino`. Now, we have to select our hardware options from the code and uncomment the right line as follows:

```
// HARDWARE OPTIONS
/*****************************************************************/
// Select your hardware here by uncommenting one line!
//#define HW__VERSION_CODE 10125 // SparkFun "9DOF Razor IMU" version "SEN-10125" (HMC5843 magnetometer)
//#define HW__VERSION_CODE 10736 // SparkFun "9DOF Razor IMU" version "SEN-10736" (HMC5883L magnetometer)
  #define HW__VERSION_CODE 14001 // SparkFun "9DoF Razor IMU M0" version "SEN-14001"
//#define HW__VERSION_CODE 10183 // SparkFun "9DOF Sensor Stick" version "SEN-10183" (HMC5843 magnetometer)
//#define HW__VERSION_CODE 10321 // SparkFun "9DOF Sensor Stick" version "SEN-10321" (HMC5843 magnetometer)
//#define HW__VERSION_CODE 10724 // SparkFun "9DOF Sensor Stick" version "SEN-10724" (HMC5883L magnetometer)
```

Hardware selection

We will compile and upload the code on the Arduino board using the IDE, just like we did in the previous section. Note that we have to choose the right processor in the Arduino IDE; in our case, it will be `Arduino/Genuino Uno`.

In the last step, we have to create the configuration file in the `razor_imu_9dof` package. We can copy the default to the `my_razor.yaml` configuration:

```
$ roscd razor_imu_9dof/config
$ cp razor.yaml my_razor.yaml
```

However, it requires the right configuration; one important configuration is setting up the port. Moreover, we can also include the calibration parameter in this file so that we have a correct orientation measurement, as shown in the following code snippet in the following screenshot:

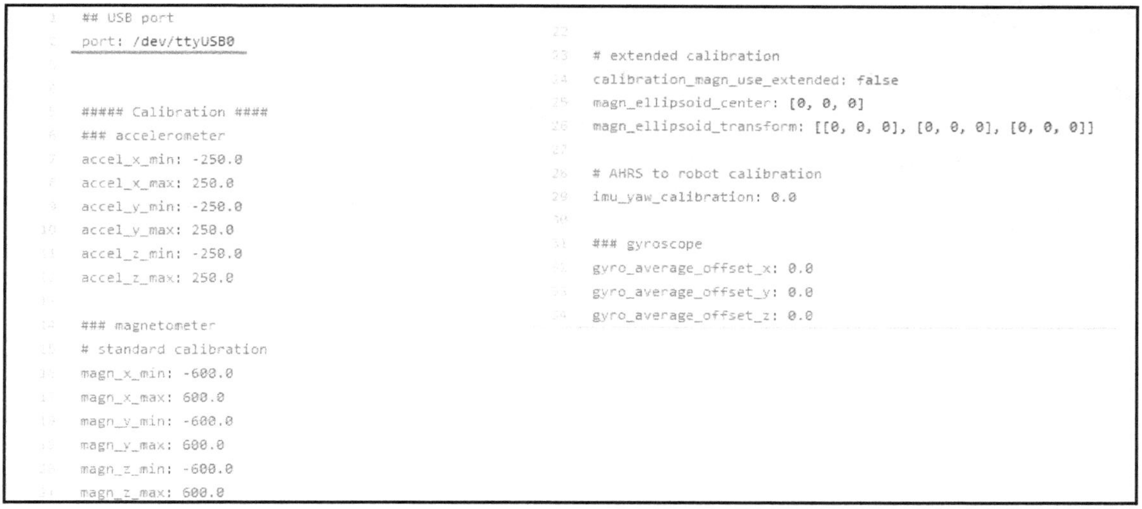

```
## USB port
port: /dev/ttyUSB0


##### Calibration ####
### accelerometer
accel_x_min: -250.0
accel_x_max: 250.0
accel_y_min: -250.0
accel_y_max: 250.0
accel_z_min: -250.0
accel_z_max: 250.0

### magnetometer
# standard calibration
magn_x_min: -600.0
magn_x_max: 600.0
magn_y_min: -600.0
magn_y_max: 600.0
magn_z_min: -600.0
magn_z_max: 600.0

# extended calibration
calibration_magn_use_extended: false
magn_ellipsoid_center: [0, 0, 0]
magn_ellipsoid_transform: [[0, 0, 0], [0, 0, 0], [0, 0, 0]]

# AHRS to robot calibration
imu_yaw_calibration: 0.0

### gyroscope
gyro_average_offset_x: 0.0
gyro_average_offset_y: 0.0
gyro_average_offset_z: 0.0
```

Configuration file for 9Dof Razer IMU M0

Finally, we can launch the `razor_imu` node using the following command:

```
$ roslaunch razor_imu_9dof razor-pub-and-display.launch
```

You may have noticed that two windows just appeared, one with 3D axis arrows with an X-Y plane, and one that shows the **Roll**, **Pitch**, and **Yaw**, as shown in the following screenshot:

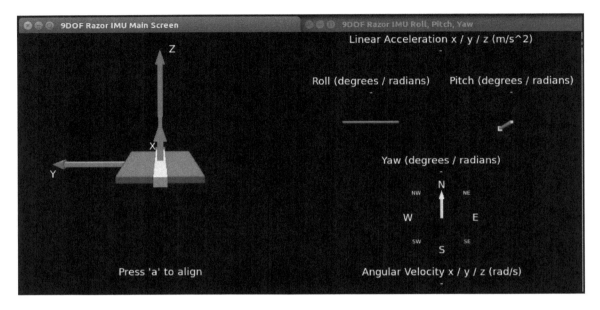

9Dof Razor IMU display

If everything is fine, we will see the topic `/imu` of `sensor_msgs/Imu` message type, using the `rostopic` list and the `type` command. The message type may be like the one listed as follows:

```
Header header

geometry_msgs/Quaternion orientation
float64[9] orientation_covariance # Row major about x, y, z axes

geometry_msgs/Vector3 angular_velocity
float64[9] angular_velocity_covariance # Row major about x, y, z axes

geometry_msgs/Vector3 linear_acceleration
float64[9] linear_acceleration_covariance # Row major x, y z
```

sensor_msgs/Imu

In the following screenshot, we can see that the parameters will change if we move our sensor:

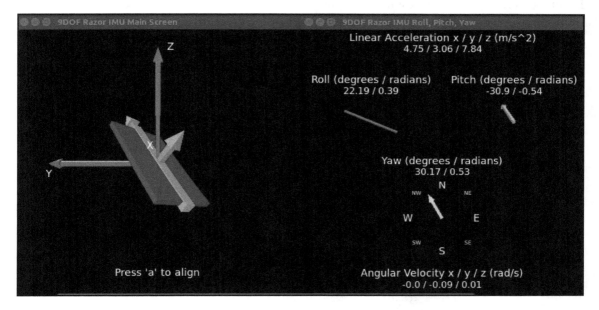

Razor IMU orientation reading

Using a GPS system – Ublox

Global Positioning System (**GPS**) is a satellite navigation system that provides location and time information in all climate conditions. The GPS can be used for navigation in planes, ships, and cars. The system gives critical abilities to the military and civilians and provides continuous real-time, three-dimensional positioning, navigation, and timing worldwide.

We can find significant differences in performance and precision, depending on the GPS receiver quality. We can have a GPS receiver that is low cost, which has errors in the range of a few meters (5 m -10 m) and is commonly used in various applications requiring less precision, such as map routing. In other cases, we can have expensive GPS devices, configured as differential GPS (DGPS). This DGPS can achieve great results in terms of precision, with location errors less than 10 cm, and that can work in **Real Time Kinematics** (**RTK**) mode.

Getting ready

In this section, we will work with the *Ublox NEO-6M GPS module*. The NEO-6M GPS module is a well-performing complete GPS receiver with a built-in 25 x 25 x 4 mm ceramic antenna, which provides a strong satellite search capability. It also provides the power and signal indicators to monitor the status of the module.

In general, a GPS module uses serial protocols to transmit data to a computer or a microcontroller, such as Arduino. Similarly, the Ublox NEO-6M GPS module has an interface of RS232 TTL, with a default baud-rate of 9600 bps and a power supply of 3-5 V. It is easy to connect to the computer with a USB adapter, as shown in the following photograph, where the NEO-6M GPS module is connected to the Linux system or PC with a USB cable:

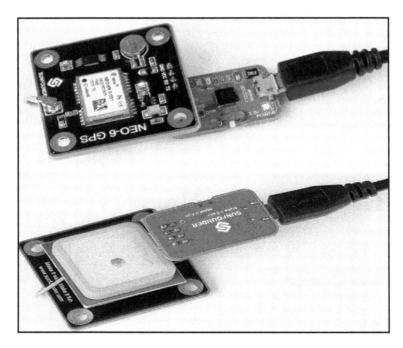

NEO-6M GPS module connection with USB

Here, we will develop and configure a setup to send the positioning data collected by the NEO-6M GPS to the ROS framework running on a Linux system or PC.

How to do it...

1. Install the NMEA GPS driver package by using the following command:

    ```
    $ sudo apt-get install ros-kinetic-nmea-gps-driver
    $ rosstack profile & rospack profile
    ```

2. Download the package from the repository and build the package from the source:

    ```
    $ cd ~/catkin_ws/src
    $ git clone git
    https://github.com/ros-drivers/nmea_navsat_driver.git
    $ cd ..
    $ catkin_make
    ```

3. To execute the GPS driver, we will run the `nmea_gpst_driver.py` script, which requires two arguments: the port name that is connected to the GPS, and the baud-rate.

     It is important to find the correct port name and supported baud-rate.

    ```
    $ sudo chmod 666 /dev/ttyUSB0
    $ rosrun nmea_gps_driver nmea_gps_driver.py _port:=/dev/ttyUSB0
    _baud:=9600
    ```

How it works...

If all the above steps are executed, we will see a topic named /fix in the topic list by executing the following command:

```
$ rostopic list
$ rostopic type /fix
```

The /fix topic has a message type of `sensor_msg/NavSatFix`.

We can echo this message to see a real stream of the data sent:

```
$ rostopic echo /fix
```

The output is shown in the following screenshot:

```
header:
  seq: 161
  stamp:
    secs: 40
    nsecs: 500000000
  frame_id: sensor
status:
  status: 0
  service: 0
latitude: -30.0602249716
longitude: -51.17391374
altitude: 9.960587315
position_covariance: [0.0025010000000000006, 0.0, 0.0, 0.0, 0.002501000000
6, 0.0, 0.0, 0.0, 0.0025010000000000006]
position_covariance_type: 2
---
```

GPS data stream

Interfacing servomotors – Dynamixel

In mobile robotics, servomotors are widely used actuators that are used to move sensors, wheels, and robotic arms.

The Dynamixel servomotor is a smart actuator system that is exclusively used for connecting joints on a robot or mechanical structure. They are designed to be modular and daisy chained on any robot or mechanical structure for powerful and flexible robotic movements.

How to do it...

1. Install the necessary packages and drivers by executing the following command:

```
$ sudo apt-get install ros-kinetic-dynamixel-motor
$ rosstack profile && rospack profile
```

2. Download the package from the repository and build the package from the source:

```
$ cd ~/catkin_ws/src
$ git clone git https://github.com/arebgun/dynamixel_motor.git
$ cd ..
$ catkin_make
```

How it works...

Once the necessary packages and drivers are installed, we will connect the dongle to the computer, as shown in the following photo, and check whether it is detected or not:

Dynamixel connected to the computer

Generally, it will create a new port with the name ttyUSBX inside the /dev/ folder. This port that means drivers have been loaded properly for the Dynamixel device, and that it is connected to the Linux system or PC.

We can launch the controller_manager node using the following command:

```
$ roslaunch dynamixel_tutorials controller_manager.launch
```

We can see that the motors have been detected by the driver on the console output. In our case, the motor with ID 4 is detected and configured:

```
process[dynamixel_manager-1]: started with pid [4968]
[INFO] [WallTime: 1259367072.683441] pan_tilt_port: Pinging motor IDs 1 through
25...
[INFO] [WallTime: 1259367074.846670] pan_tilt_port: Found 1 motors - 1 AX-12 [4]
, initialization complete.
```

Dynamixel driver initialization

After launching the `controller_manager.launch` file, we will notice a list of topics, as follows:

```
$ rostopic list
/diagnostics
/motor_states/pan_tilt_port
/rosout
/rosout_agg
```

Control manager topic list

We could echo the `/motor_states/pan_tilt_port` topic, which will show the state of all the motors connected to the Linux system; in our case, it is only the motor with the ID of 4. Since we cannot control the motors with these topics, we will need to launch the `dynamixel_controllers` node using the following `launch` script:

$ **roslaunch dynamixel_tutorials controller_spawner.launch**

Now, two new topics have been added to the list, which is `/tilt_controller/command` and `/tilt_controller/state`.

We are going to use the `/tilt_controller/command` topic of the `std_msgs/Float64` message type to control the motor movement as follows:

$ **rostopic pub /tilt_controller/command std_msgs/Float64 -- 0.4**

Where `Float64` is a variable that is used to move the motor to a position measured in radians. Once the command is executed, we will see that the motor is moving, and it will stop at `0.4` radians.

Moreover, we can also connect the daisy chain of Dynamixel motors to the Linux system, as shown in the following diagram:

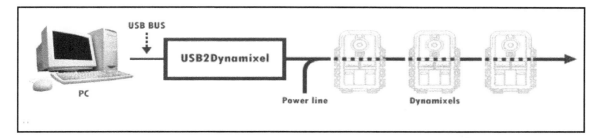

Daisy chain of Dynamixel motors

As mentioned previously, during initialization of the driver, each motor gets its own ID that will be used for future communication.

Using a Laser Rangefinder – Hokuyo

Range sensing is a crucial element of any environment and obstacle mapping system for mobile robotics applications. Sensors that are suitable for the environment and obstacle mapping include 2D Ladar and 3D Ladar. However, a cost-effective approach for the environment and obstacle mapping is to mount a 2D laser scanner that is aimed forward and downward on the front end of a mobile robot.

Getting ready

In this section, we will learn how to use a low-cost model of Lidar which is widely used in robotics applications, such as *Hokuyo URG-04LX*. The Hokuyo Laser Range Finder is similar in function to the *Sick Laser Range Finder*, which has been the de-facto standard range sensor for mobile robot obstacle avoidance and mapping applications for the last decade. The Sick Laser Range Finder is of relatively large size, weight, and power consumption, which means you can only use it on relatively large mobile robots. On the other hand, the Hokuyo Laser Range Finder is substantially smaller, lighter, and consumes less power, and is, therefore, more suitable for small, mobile robots.

We can obtain more information about the Hokuyo Laser Range Finder at http://www.hokuyo-aut.jp. The Hokuyo Laser Range Finder can be used to navigate and build maps in real time.

How to do it...

1. To start using the Hokuyo URG-04LX-UG01, we have to install the drivers.
2. Download the package from the repository and build the package from the source since ROS distribution does not contain the pre-built binary package:

```
$ cd ~/catkin_ws/src
$ git clone git clone
https://github.com/ros-drivers/driver_common.git
$ git clone https://github.com/ros-drivers/hokuyo_node.git
$ cd ..
$ catkin_make
```

How it works...

Once the necessary packages and drivers are built and installed, we will connect the Hokuyo URG-04LX-UG01 to the computer using a USB cable and check whether it is detected or not:

```
$ ls -l /dev/ttyACM0
```

When the laser is connected, the system detects it, and the preceding command will show the following output:

```
crw-rw---- 1 root dialout 166, 0 May 14 12:06 /dev/ttyACM0
```

 Note that our case system has created the node /dev/ttyACM0; it could be /dev/ttyACMX for any other system.

In some cases, we have to reconfigure the laser device node to provide access permission. By default, it only has root permission:

```
$ sudo chmod a+rw /dev/ttyACM0
```

Once everything is good to go, we will switch on the Hokuyo URG-04LX-UG01 device and start roscore in one terminal, and in another terminal, execute the following command:

```
$ rosrun hokuyo_node hokuyo_node
```

The preceding command will show the following output if hokuyo_node started correctly:

```
[ INFO] [1458876560.194647219]: Connected to device with ID: H1000589
```

We can use the `rostopic` list to see all the topics created by `hokuyo_node`, which will show the following output:

```
/diagnostics
/hokuyo_node/parameter_descriptions
/hokuyo_node/parameter_updates
/rosout
/rosout_agg
/scan
```

rostopic for hokuyo_node

Where the `/scan` topic of message type `sensor_msgs/LaserScan` is used to publish the information about the laser scan. We can see how the laser works and what data it is sending by using the `rostopic` command:

$ rostopic echo /scan

This will show the following outputs, as depicted in the following diagram. It is difficult to understand the laser scan data for humans. We can represent the data in a more friendly and graphical way by using `rviz` visualization tools. We will learn more about configuring the `rviz` to visualize the laser scan in the next chapter in more detail:

Laser scanning data

Working with the Kinect sensor to view objects in 3D

The innovative technology behind Kinect is a combination of hardware and software contained within the Kinect sensor accessory that can be added to any existing Xbox 360 or computer system. There are three hardware innovations working together within the Kinect sensor:

- **Color VGA video camera**: This video camera aids in facial recognition and other detection features by detecting three color components—red, green, and blue, which is also known as the "RGB camera".
- **Depth sensor**: An infrared projector and a monochrome CMOS sensor work together to "see" the indoor environment in 3D, regardless of the lighting conditions.
- **Multi-array microphone**: This is an array of four microphones that can isolate the voices of the players from the noise in the indoor environment. This allows the player to be a few feet away from the microphone and still use voice controls.

We can find the Microsoft Kinect on the Microsoft product website (https://msdn. microsoft.com/en-us/library/hh438998.aspx). A further look at the technical specifications for Kinect reveals that both the video and depth sensor cameras have a 640 x 480-pixel resolution and run at 30 FPS. The specifications also suggest that depth range works reasonable well in the range of 6 feet (1.5 meters).

Getting ready

Here, we will use two of these sensors: the RGB camera and the depth sensor. However, with the latest version of ROS packages, we can even use all three.

How to do it...

1. Install the packages and drivers for Microsoft Kinect, as follows:

```
$ sudo apt-get install ros-kinetic-openni-camera
$ rosstack profile && rospack profile
```

2. Download the `openni_camera` package from the repository and build the package from the source:

```
$ cd ~/catkin_ws/src
$ git clone https://github.com/ros-drivers/openni_camera.git
$ cd ..
$ catkin_make
```

How it works...

Once the packages and drivers have been built and installed, we can plug the Microsoft Kinect sensor into the Linux system or PC, and then run the following command to run the nodes to start using the sensors:

```
$ rosrun openni_camera openni_node
$ roslaunch openni_launch openni.launch
```

`roscore` must be running before executing the preceding command. No errors will occur if all works to plan.

Next, we will learn how to use the Kinect sensors. List the `rostopics` created by using the preceding launch command when the sensor driver and nodes start. Although the `rostopic` list shows that many topics have been created, the following diagram shows some of the topics that are important to discuss:

```
...
/camera/rgb/image_color
/camera/rgb/image_mono
/camera/rgb/image_raw
/camera/rgb/image_rect
/camera/rgb/image_rect_color
...
```

rostopic list for Kinect sensor

To view the image from the sensor, we will have to use the `image_view` package as follows:

```
$ rosrun image_view image_view image:=/camera/rgb/image_color
```

Here, you may have noticed that we have to rename (remap) the image topic to /camera/rgb/image_color using the parameter's image.

Well! If everything works fine, a new window will appear that shows the diagram from Kinect. Similarly, we can view the image from the depth sensor by changing the topic in the previous command line:

```
$ rosrun image_view image_view image:=/camera/depth/image
```

Good luck! We will then see an image similar to the following screenshot:

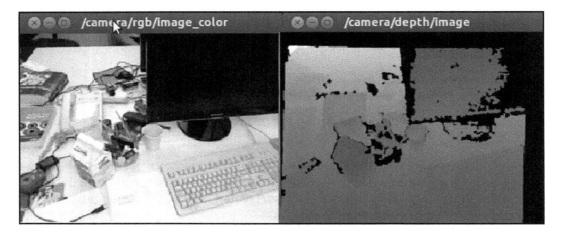

Kinect: RGB image and depth image

Another important topic is the /camera/depth/points, which publishes the point cloud data. The point cloud data is a 3D representation of the depth image. However, we can view the point cloud data in RViz, a 3D visualization tools provided by the ROS. The following screenshot shows the visualization of point cloud in RViz. Notice that the pointcloud2 topic is set to /camera/depth/points in the RViz display panel:

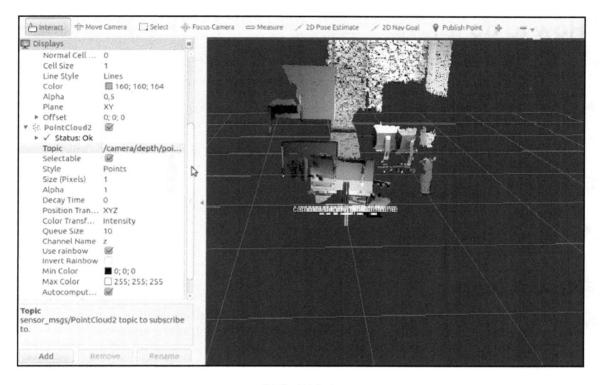

PointCloud data in rviz

Using a joystick or a gamepad in ROS

A joystick is an input device that can be used for controlling the movement of the cursor or a pointer on a computer device. However, with this device, we can perform or control a wide range of actions.

In this section, we will discuss how to interface the joystick device to the ROS framework and control a real or simulated robot. A wide range of joystick devices is available for use in several applications.

How to do it...

1. Install the necessary packages and driver, before working with the joystick.
2. To install these packages in Ubuntu, we have to execute the following command:

```
$ sudo apt-get install ros-kinetic-joystick-drivers
$ rosstack profile & rospack profile
```

3. Download the `joystick_drivers` package from the repository and build the package from the source:

```
$ cd ~/catkin_ws/src
$ git clone https://github.com/ros-drivers/joystick_drivers.git
$ cd ..
$ catkin_make
```

How it works...

After installing the packages, we will connect our joystick to the computer system and check whether the joystick is recognized or not by using the following command:

```
$ ls /dev/input/
```

This will produce an output showing that all device nodes in the system correspond to input devices, as shown in the following screenshot:

```
by-id event0 event2 event4 event6 event8 js0 mouse0
by-path event1 event3 event5 event7 event9 mice
```

System input devices

You will find that the `js0` device node was created by the driver when the joystick was plugged into the computer system. Next, we can use the `jstest` command so we can check whether it is working or not:

```
$ sudo jstest /dev/input/js0
```

The output is shown in the following screenshot:

```
Axes: 0: 0 1: 0 2: 0 Buttons: 0:off 1:off 2:off 3:off 4:off 5:off 6:off 7:off
8:off 9:off 10:off
```

jstest result for the joystick

This shows the configuration of our joystick, the Logitech F710, which has 8 axes and 11 buttons. As we move the joystick, the values change.

We have seen that our joystick is working fine. Next, we need to verify its interface with the ROS framework by using the following command:

```
$ rosrun joy joy_node
```

We will see the following output if the `joy_node` works without any errors:

```
[ INFO] [1357571588.441808789]: Opened joystick: /dev/input/js0. deadzone_: 0.
050000.
```

joy_node starup messages

The `rostopic list` command shows that the `/joy` topic of type `sensor_msgs/Joy` message is created by `joy_node`. To see the messages sent by the node, we can use the following command:

```
$ rostopic echo /joy
```

We will see an output similar to what is shown in the following screenshot:

```
joy topic output                                          ≡ ◇ ≡ ↔ 🔳 ❐ Vim
---
header:
  seq: 429
  stamp:
    secs: 1415227355
    nsecs: 833352850
  frame_id: ''
axes: [0.19235174357891083, -0.0, 1.0, -0.04268254339694977, -0.048208002001047134, 1.0
buttons: [0, 0, 0, 0, 0, 0, 0, 0, 0, 0, 0, 0, 0, 0, 0]
---
```

topic joy outputs

We can also write a node that subscribes to the joystick topic and generate moving commands to move the robot model.

ROS Modeling and Simulation

6

In this chapter, we will discuss the following recipes:

- Understanding robot modeling using URDF
- Understanding robot modeling using xacro
- Understanding the joint state publisher and the robot state publisher
- Understanding the Gazebo architecture and interface with ROS

Introduction

In the previous chapter, we learned how to use, configure, and explore some specific sensors and actuators that are broadly used in the world of robotics. However, working and programming with the real hardware and robot would be more exciting and compelling for a real-world application, but it is not possible and reasonable for every robotic developer to use a real robot during development. Hence, we will learn how to simulate the physical world and robot.

Simulators are great development tools when we have limited access to a real robot and hardware, which were developed for validating and testing the behavior of algorithms before applying them on a real robot during the development phase.

In this chapter, we will learn how to develop a 3D model of a robot that is required for a real-world application. This includes how to create joints and add texture to a model, as well as how to use a ROS node to control the robot.

We will also introduce Gazebo, a simulator framework that is widely used by the ROS community and already supports several real robots in the simulation. We can use Gazebo to load the 3D models of our robots and simulate their perception and control them in a virtual world. Moreover, we will learn how to use parts of other robots designed by the community, particularly grippers and sensors, such as a laser range finder and a camera.

Understanding robot modeling using URDF

The ROS framework provides the **Unified Robot Description Format** (URDF), which is an XML file format that describes a robot, its geometry, its parts, its joints, physics, and so on. Whenever we use a 3D robot on ROS, for example, a mobile robot (Turtlebot), the humanoid robot (PR2), and some types of aerial robot, a set of several URDF files must be associated with them.

In the following sections, we will learn how to use the URDF file format to create a 3D model of a real-world robot for development.

Getting ready

We are going to build a mobile robot with four wheels and an arm robot with a gripper. Moreover, we can create each example program in the ROS package in our workspace individually or copy the `chapter6_tutorials` packages from GitHub (https://github.com/kbipin/Robot-Operating-System-Cookbook), which has two packages called `robot_description` and `robot_gazebo`.

How it works...

Build the base of the robot with four wheels by creating a file in the `chapter6_tutorials/robot_description/urdf` folder with the name `mobile_robot.urdf`, and add the following code:

```
<?xml version="1.0"?>
<robot name="mobile">
    <link name="base_link">
        <visual>
            <geometry>
                <box size="0.2 .3 .1"/>
            </geometry>
            <origin rpy="0 0 0" xyz="0 0 0.05"/>
            <material name="yellow">
                <color rgba="255 255 0 1"/>
            </material>
        </visual>
        <collision>
            <geometry>
                <box size="0.2 .3 0.1"/>
            </geometry>
```

```
            </collision>
            <inertial>
                <mass value="100"/>
                <inertia ixx="1.0" ixy="0.0" ixz="0.0" iyy="1.0" iyz="0.0"
izz="1.0"/>
            </inertial>
        </link>
        <link name="wheel_1">
            <visual>
                <geometry>
                    <cylinder length="0.05" radius="0.05"/>
                </geometry>
                <origin rpy="0 1.5 0" xyz="0.1 0.1 0"/>
                <material name="black">
                    <color rgba="0 0 0 1"/>
                </material>
            </visual>
            <collision>
                <geometry>
                    <cylinder length="0.05" radius="0.05"/>
                </geometry>
            </collision>
            <inertial>
                <mass value="10"/>
                <inertia ixx="1.0" ixy="0.0" ixz="0.0" iyy="1.0" iyz="0.0"
izz="1.0"/>
            </inertial>
        </link>
        <link name="wheel_2">
            <visual>
                <geometry>
                    <cylinder length="0.05" radius="0.05"/>
                </geometry>
                <origin rpy="0 1.5 0" xyz="-0.1 0.1 0"/>
                <material name="black"/>
            </visual>
            <collision>
                <geometry>
                    <cylinder length="0.05" radius="0.05"/>
                </geometry>
            </collision>
            <inertial>
                <mass value="10"/>
                <inertia ixx="1.0" ixy="0.0" ixz="0.0" iyy="1.0" iyz="0.0"
izz="1.0"/>
            </inertial>
        </link>
        <link name="wheel_3">
```

```
            <visual>
                    <geometry>
                            <cylinder length="0.05" radius="0.05"/>
                    </geometry>
                    <origin rpy="0 1.5 0" xyz="0.1 -0.1 0"/>
                    <material name="black"/>
            </visual>
            <collision>
                    <geometry>
                            <cylinder length="0.05" radius="0.05"/>
                    </geometry>
            </collision>
            <inertial>
                    <mass value="10"/>
                    <inertia ixx="1.0" ixy="0.0" ixz="0.0" iyy="1.0" iyz="0.0"
izz="1.0"/>
            </inertial>
    </link>
    <link name="wheel_4">
            <visual>
                    <geometry>
                            <cylinder length="0.05" radius="0.05"/>
                    </geometry>
                    <origin rpy="0 1.5 0" xyz="-0.1 -0.1 0"/>
                    <material name="black"/>
            </visual>
            <collision>
                    <geometry>
                            <cylinder length="0.05" radius="0.05"/>
                    </geometry>
            </collision>
            <inertial>
                    <mass value="10"/>
                    <inertia ixx="1.0" ixy="0.0" ixz="0.0" iyy="1.0" iyz="0.0"
izz="1.0"/>
            </inertial>
    </link>
    <joint name="base_to_wheel1" type="continuous">
            <parent link="base_link"/>
            <child link="wheel_1"/>
            <origin xyz="0 0 0"/>
    </joint>
    <joint name="base_to_wheel2" type="continuous">
            <parent link="base_link"/>
            <child link="wheel_2"/>
            <origin xyz="0 0 0"/>
    </joint>
    <joint name="base_to_wheel3" type="continuous">
```

```
        <parent link="base_link"/>
        <child link="wheel_3"/>
        <origin xyz="0 0 0"/>
    </joint>
    <joint name="base_to_wheel4" type="continuous">
        <parent link="base_link"/>
        <child link="wheel_4"/>
        <origin xyz="0 0 0"/>
    </joint>
</robot>
```

The preceding URDF code has an XML file format, where indentation is not necessary, but recommended. We can use a reasonably advanced editor with the required plugins or configuration such as an appropriate `.vimrc` file in vim or the atom editor by default:

1. **URDF file format**:
   - You may have noticed that in the code, there are two primary elements that define the geometry of a robot, links, and joints, where the first link has the name `base_link`, which must be unique to the file:

```
<link name="base_link">
    <visual>
        <geometry>
            <box size="0.2 .3 .1"/>
        </geometry>
        <origin rpy="0 0 0" xyz="0 0 0.05"/>
        <material name="yellow">
            <color rgba="255 255 0 1"/>
        </material>
    </visual>
    ...
</link>
```

   - The `visual` element in the preceding code defines what will be displayed on the simulator, where we can define the geometry—cylinder, box, sphere, mesh, the material—color, texture, and the origin. Then, we have to code the joint, as follows:

```
<joint name="base_to_wheel1" type="continuous">
    <parent link="base_link"/>
    <child link="wheel_1"/>
    <origin xyz="0 0 0"/>
</joint>
```

- Similarly, in the `joint` field which must be unique as well, define the type of joint, which is going to be `fixed`, `revolute`, `continuous`, `floating`, or `planar`, along with the parent and child relationship between the joints. Here, `base_link` is the unique parent of `wheel_1`.

- Moreover, we can check the correctness of the syntax in the previously created URDF file by using the `check_urdf` command tool:

  ```
  $ check_urdf mobile_robot.urdf
  ```

- The expected output is shown in the following screenshot:

```
kumar@kumar-Inspiron-5437:~/catkin_ws-01/src/chapter6_tutorials/robot_description/urdf$ check_urdf mobile_robot.urdf
robot name is: mobile robot
---------- Successfully Parsed XML ---------------
root Link: base_link has 4 child(ren)
    child(1):  wheel_1
    child(2):  wheel_2
    child(3):  wheel_3
    child(4):  wheel_4
```

URDF syntax

- We can view the architecture diagram graphically using the `urdf_to_graphiz` command tool:

  ```
  $ urdf_to_graphiz mobile_robot.urdf
  ```

- This command tool will generate two files, which are `mobile.pdf` and `mobile.gv`. We can open the file with `evince` as follows:

  ```
  $ evince mobile.pdf
  ```

The following shows the architecture diagram received as output:

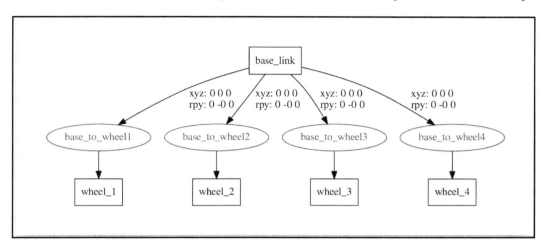

Architecture diagram

2. **3D Model on Rviz**:

   - We have discussed how to create a 3D model robot. From here, we can view the 3D model simulation using `rviz`. We will create the `simulation.launch` file in the `robot_description/launch` folder and add the following code. Alternatively, we can get the file from GitHub (`https://github.com/kbipin/Robot-Operating-System-Cookbook`):

```xml
<?xml version="1.0"?>
<launch>
    <arg name="model" />
    <arg name="gui" default="False" />
    <param name="robot_description" textfile="$(arg
model)" />
    <param name="use_gui" value="$(arg gui)"/>
    <node name="joint_state_publisher"
pkg="joint_state_publisher"
type="joint_state_publisher" ></node>
    <node name="robot_state_publisher"
pkg="robot_state_publisher" type="state_publisher" />
    <node name="rviz" pkg="rviz" type="rviz" args="-d
$(find robot_description)/robot.rviz" />
</launch>
```

- Then, we can launch the robot model using the following command:

```
$ roslaunch robot_description simulation.launch
model:="'rospack find
robot_description'/urdf/mobile_robot.urdf"
```

- Thereupon, we can view the 3D model of the mobile robot, as shown in the following screenshot:

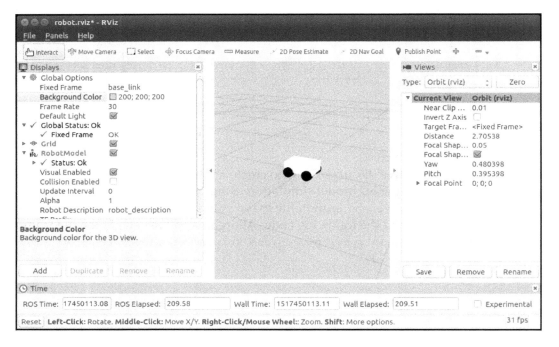

3D model of the robot

- Similarly, we will design a 3D model of a robot manipulator, which will have a base and an articulated arm with a gripper. We can get the complete model from the `chapter6_tutorials/robot_description/urdf/arm_robot.urdf` file in GitHub. Then, we can launch the robot model by using the following command:

```
$ roslaunch robot_description display.launch
model:="'rospack find
robot1_description'/urdf/arm_robot.urdf"
```

- We can view the 3D model of the robot manipulator, as shown in the following screenshot:

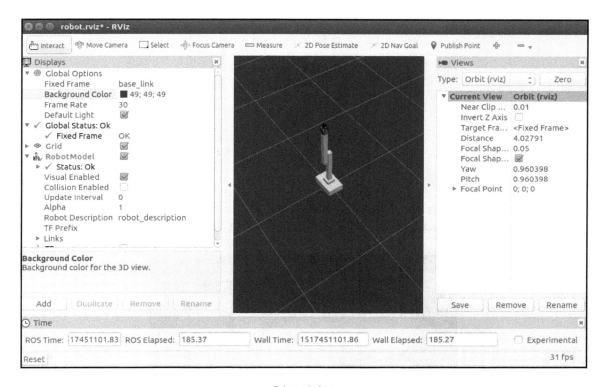

Robot manipulator

3. **Meshes with the 3D model**:
    - We would like to have a realistic look for our model or a more detailed design, rather than basic geometric blocks. This is possible using meshes, which might be custom generated or meshes of other models that are available from the community.
    - In the case of the 3D model of the robot manipulator, which we discussed previously, we will use PR2's gripper. The following code snippet describes how to use it:

```
<link name="left_gripper">
        <visual>
                <origin rpy="0 0 0" xyz="0 0 0"/>
                <geometry>
                        <mesh
```

```
filename="package://urdf_tutorial/meshes/l_finger.dae"
/>
                </geometry>
        </visual>
        <collision>
                <geometry>
                        <box size="0.1 .1 .1"/>
                </geometry>
        </collision>
        <inertial>
                <mass value="1"/>
                <inertia ixx="1.0" ixy="0.0" ixz="0.0"
iyy="1.0" iyz="0.0" izz="1.0"/>
        </inertial>
</link>
```

- However, this resembles the sample link that we used before, but in the geometry section, we added the mesh. Consequently, you can see the result in the following diagram:

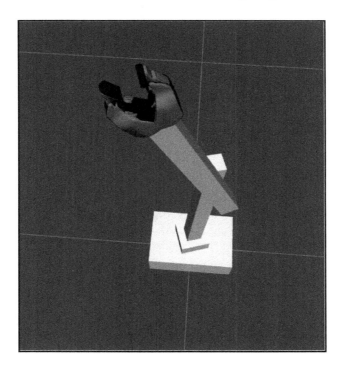

Mesh model

4. **Control and motion**:

- The 3D model of the robot can have control and motion according to the joints described in the URDF file. The most used type of joint is the revolute joint, in the case of the arm robot.
- You may have noticed it being used on `arm_1_to_arm_base` in the following code:

```
<joint name="arm_1_to_arm_base" type="revolute">
        <parent link="arm_base"/>
        <child link="arm_1"/>
        <axis xyz="1 0 0"/>
        <origin xyz="0 0 0.15"/>
        <limit effort ="1000.0" lower="-1.0"
upper="1.0" velocity="0.5"/>
</joint>
```

- The revolute joints have strict limits, which are fixed using the `<limit effort ="1000.0" lower="-1.0" upper="1.0" velocity="0.5"/>` line in the URDF file. We can also select the axis of rotation using `<axis xyz="1 0 0">`. We can use the `<limit>` tag to define the set of attributes—`effort` defines the maximum force supported by the joint, `lower` is used to assign a lower limit, for example, radian for revolute joints and meters for prismatic joints, `upper` is used for the upper limit, and `velocity` is used for enforcing the maximum joint velocity.
- We can confirm the axis and limits of the joints' attributes by running `rviz` with the `join_state_publisher` graphically, as follows:

```
$ roslaunch robot_description simulation.launch
model:="'rospack find
robot_description'/urdf/arm_robot.urdf" gui:=true
```

- The GUI will pop up, as shown in the following screenshot:

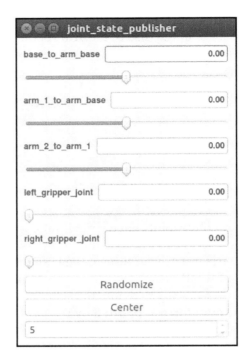

Joint state publisher

- Similarly, the mobile robot model uses a continuous-type joint, as shown in the following code snippet:

```
<joint name="base_to_wheel1" type="continuous">
        <parent link="base_link"/>
        <child link="wheel_1"/>
        <origin xyz="0 0 0"/>
</joint>
```

5. **Physical properties**:
   - To simulate the 3D robot model in Gazebo or any other simulation software which has a physics engine, the physical and collision properties of the model must be defined in the URDF files. In other words, the model must define its geometry to calculate the possible collisions and the weight in order to measure its inertia.

- It is essential that all links defined in the model file have these parameters, otherwise the robot model will not be simulated by the physics engine. However, calculating the collisions among meshes is computationally complex, and therefore the simplified geometry of a mesh model can be used rather than the actual mesh.

- In the following code, you may notice the physical and collision properties of `link` with the name `wheel_1` in `mobile_robot.urdf` for the mobile robot:

```
<link name="wheel_1">
        <visual>
                <geometry>
                        <cylinder length="0.05"
radius="0.05"/>
                </geometry>
                <origin rpy="0 1.5 0" xyz="0.1 0.1 0"/>
                <material name="black">
                        <color rgba="0 0 0 1"/>
                </material>
        </visual>
        <collision>
                <geometry>
                        <cylinder length="0.05"
radius="0.05"/>
                </geometry>
        </collision>
        <inertial>
                <mass value="10"/>
                <inertia ixx="1.0" ixy="0.0" ixz="0.0"
iyy="1.0" iyz="0.0" izz="1.0"/>
        </inertial>
</link>
```

Similarly, collision and inertial elements are defined for all the links in both the mobile and arm robot, as discussed previously, since if we do not do so, Gazebo will not simulate the model.

We can find a complete file with all the parameters in `mobile_robot.urdf` and `arm_robot.urdf` at `robot_description/urdf/` at GitHub (`https://github.com/kbipin/Robot-Operating-System-Cookbook`).

Understanding robot modeling using xacro

In this section, we will discuss `xacro`, XML macros which define the URDF files in a compact form and makes their readability and maintainability easier. Moreover, this new format allows us to develop the robot model in a modular structure, which can be reused in complex design.

Getting ready

Once again, we are going to build a mobile robot with four wheels and an arm robot with a gripper, like we discussed in the previous section, using `xacro` instead of the URDF file format. Moreover, we can create each example program in the ROS package in our workspace individually or copy them from the `chapter6_tutorials` packages in GitHub, which are `arm_robot.xacro` and `mobile_robot.xacro`, respectively.

How it works...

To begin with `xacro`, we have to specify a namespace so that the file is parsed properly, as shown in the following code snippet:

```
<?xml version="1.0"?>
<robot xmlns:xacro="http://www.ros.org/wiki/xacro" name="mobile">
```

We must remember that the file must have an `xacro` extension in place of URDF:

1. **Constants**:
   - We can use `xacro` to declare constant values, just like we do in any other programming language. For example, in the case of the mobile robot which has four wheels with the same values for length and radius, this can be declared and defined as follows:

     ```
     <xacro:property name="length_wheel" value="0.05" />
     <xacro:property name="radius_wheel" value="0.05" />
     ```

   - These values can be used in the following ways, later in the code:

     ```
     ${name_of_variable}:
     <cylinder length="${length_wheel}"
     radius="${radius_wheel}"/>
     ```

2. **Mathematics**:

- We can also develop complex mathematical expressions in the ${}
  construct using basic operations, including +, -, *, /, unary minus, and
  parentheses. However, exponential and modulus are not supported:

```
<cylinder radius="${wheel_diameter/2}" length="0.1"/>
<origin xyz="${reflection*(width+0.04)} 0 0.25" />
```

Thus, the parameterized design is possible by using mathematics in
xacro.

3. **Macros**:

- One of the most useful elements of the xacro package would be
  macros. For example, in our 3D robot model design description, we
  will use the following macro for inertial:

```
<xacro:macro name="default_inertial" params="mass">
    <inertial>
        <mass value="${mass}" />
        <inertia ixx="1.0" ixy="0.0" ixz="0.0"
          iyy="1.0" iyz="0.0"
          izz="1.0" />
        </inertial>
</xacro:macro>
```

- We can compare the size of mobile_robot.urdf and
  arm_robot.urdf with mobile_robot.xacro and
  arm_robot.xacro, respectively, where we will have to eliminate
  at least 30 duplicate lines without much effort; however, it is
  possible to reduce this further by using a structure or modular
  design with more macros and variables.
- We can convert the .xacro file to .urdf, which can be used with
  rviz as in the previous section by using the following command:

```
$ rosrun xacro xacro.py "'rospack find
robot_description'/urdf/mobile_robot.xacro" >
"'rospack find
robot_description'/urdf/mobile_robot_processed.urdf"
$ rosrun xacro xacro.py "'rospack find
robot_description'/urdf/arm_robot.xacro" > "'rospack
find robot_description'/urdf/arm_robot_processed.urdf"
```

Understanding the joint state publisher and the robot state publisher

At this moment in time, we have 3D models of robots which can be visualized in `rviz`. Consequently, in this section, we will learn how to control and move the robot model using the ROS node. The ROS framework provides great tools to control the robots, such as the `ros_control` package. Similarly, other primary tools to control the robots are the `joint_state_publisher` and `robot_state_publisher` packages.

Getting ready

Like we did in the previous section, we can create each example program in the ROS packages in our workspace individually in the `robot_description/src` folder with the names `mobile_state_publisher.cpp` and `arm_state_publisher.cpp`, or copy them from the `chapter6_tutorials` packages on GitHub.

The `mobile_state_publisher.cpp` package should include the following code:

```cpp
#include <string>
#include <ros/ros.h>
#include <sensor_msgs/JointState.h>
#include <tf/transform_broadcaster.h>
int main(int argc, char** argv) {
    ros::init(argc, argv, "mobile_state_publisher");
    ros::NodeHandle n;
    tf::TransformBroadcaster broadcaster;
    ros::Publisher joint_pub =
n.advertise<sensor_msgs::JointState>("joint_states", 1);
    ros::Rate loop_rate(30);

    const double degree = M_PI/180;

    // robot state
    double angle= 0;

    // message declarations
    geometry_msgs::TransformStamped odom_trans;
    sensor_msgs::JointState joint_state;
    odom_trans.header.frame_id = "odom";
    odom_trans.child_frame_id = "base_link";

    while (ros::ok()) {
```

```
        //update joint_state
        joint_state.header.stamp = ros::Time::now();
        joint_state.name.resize(4);
        joint_state.position.resize(4);
        joint_state.name[0] ="base_to_wheel1";
        joint_state.position[0] = 0;
        joint_state.name[1] ="base_to_wheel2";
        joint_state.position[1] = 0;
        joint_state.name[2] ="base_to_wheel3";
        joint_state.position[2] = 0;
        joint_state.name[3] ="base_to_wheel4";
        joint_state.position[3] = 0;

        // update transform
        // (moving in a circle with radius)
        odom_trans.header.stamp = ros::Time::now();
        odom_trans.transform.translation.x = cos(angle);
        odom_trans.transform.translation.y = sin(angle);
        odom_trans.transform.translation.z = 0.0;
        odom_trans.transform.rotation =
tf::createQuaternionMsgFromYaw(angle);

        //send the joint state and transform
        joint_pub.publish(joint_state);
        broadcaster.sendTransform(odom_trans);


        // Create new robot state
        angle += degree/4;

        // This will adjust as needed per iteration
        loop_rate.sleep();
    }
    return 0;
}
```

Similarly, the following code should be in `arm_state_publisher.cpp`:

```
#include <string>
#include <ros/ros.h>
#include <sensor_msgs/JointState.h>
//#include <tf/transform_broadcaster.h>

int main(int argc, char** argv) {

    ros::init(argc, argv, "arm_state_publisher");
    ros::NodeHandle n;
    ros::Publisher joint_pub =
```

```
n.advertise<sensor_msgs::JointState>("joint_states", 1);
    ros::Rate loop_rate(30);

    const double degree = M_PI/180;

    // robot state
    double inc= 0.005, base_arm_inc= 0.005, arm1_armbase_inc= 0.005,
arm2_arm1_inc= 0.005, gripper_inc= 0.005, tip_inc= 0.005;
    double angle= 0 ,base_arm = 0, arm1_armbase = 0, arm2_arm1 = 0, gripper
= 0, tip = 0;

    // message declarations
    sensor_msgs::JointState joint_state;

    while (ros::ok()) {
        //update joint_state
        joint_state.header.stamp = ros::Time::now();
        joint_state.header.frame_id = "base_link";
        joint_state.name.resize(7);
        joint_state.position.resize(7);
        joint_state.name[0] ="base_to_arm_base";
        joint_state.position[0] = base_arm;
        joint_state.name[1] ="arm_1_to_arm_base";
        joint_state.position[1] = arm1_armbase;
        joint_state.name[2] ="arm_2_to_arm_1";
        joint_state.position[2] = arm2_arm1;
        joint_state.name[3] ="left_gripper_joint";
        joint_state.position[3] = gripper;
        joint_state.name[4] ="left_tip_joint";
        joint_state.position[4] = tip;
        joint_state.name[5] ="right_gripper_joint";
        joint_state.position[5] = gripper;
        joint_state.name[6] ="right_tip_joint";
        joint_state.position[6] = tip;


        //send the joint state and transform
        joint_pub.publish(joint_state);
        // Create new robot state
        arm2_arm1 += arm2_arm1_inc;
        if (arm2_arm1<-1.5 || arm2_arm1>1.5) arm2_arm1_inc *= -1;
        arm1_armbase += arm1_armbase_inc;
        if (arm1_armbase>1.2 || arm1_armbase<-1.0) arm1_armbase_inc *= -1;
        base_arm += base_arm_inc;
        if (base_arm>1. || base_arm<-1.0) base_arm_inc *= -1;
        gripper += gripper_inc;
        if (gripper<0 || gripper>1) gripper_inc *= -1;
        angle += degree/4;
```

```
        // This will adjust as needed per iteration
        loop_rate.sleep();
    }
    return 0;
}
```

In the following section, we will discuss how the preceding code is able to control and move the robot.

How it works...

In the case of the mobile robot, we will begin our discussion with the concept of tf frames and their relationship and transformation. The most used tf frames in ROS are map, odom, and base_link, where the tf frame map is a world-fixed frame which defines the long-term global reference, and the odom frame provides an accurate short-term local reference. The base_link frame is rigidly attached to the center of the curvature of the mobile robot's base. Usually, the relationship among these frames can be characterized as map | odom | base_link. In most general cases, when the map frame is not defined, the world frame is used as a global reference:

Accordingly, in our basic control node mobile_state_publisher.cpp, we will define an odom frame and all the transformations will be referred to that odom frame. In our design, all the links are children of base_link, so all of the frames are automatically liked to the odom frame. We can refer to the following code where these relationships are defined:

```
geometry_msgs::TransformStamped odom_trans;
sensor_msgs::JointState joint_state;
odom_trans.header.frame_id = "odom";
odom_trans.child_frame_id = "base_link";
```

Moreover, we will create a topic named joint_state to control all the joints of the model, where the joint_state message field holds data to describe the state of every torque-controlled joint. Since our 3D arm robot model has seven joints, we will create a joint_state message with seven elements:

```
joint_state.header.stamp = ros::Time::now();
joint_state.header.frame_id = "base_link";
joint_state.name.resize(7);
joint_state.position.resize(7);
joint_state.name[0] ="base_to_arm_base";
joint_state.position[0] = base_arm;
joint_state.name[1] ="arm_1_to_arm_base";
```

```
joint_state.position[1] = arm1_armbase;
joint_state.name[2] ="arm_2_to_arm_1";
joint_state.position[2] = arm2_arm1;
joint_state.name[3] ="left_gripper_joint";
joint_state.position[3] = gripper;
joint_state.name[4] ="left_tip_joint";
joint_state.position[4] = tip;
joint_state.name[5] ="right_gripper_joint";
joint_state.position[5] = gripper;
joint_state.name[6] ="right_tip_joint";
joint_state.position[6] = tip;
```

In the case of the mobile robot, we would like to move the robot in a circle. The following code defines the coordinates in the message fields:

```
odom_trans.header.stamp = ros::Time::now();
odom_trans.transform.translation.x = cos(angle)*1;
odom_trans.transform.translation.y = sin(angle)*1;
odom_trans.transform.translation.z = 0.0;
odom_trans.transform.rotation = tf::createQuaternionMsgFromYaw(angle);
```

Finally, we will publish the new state of our robot in every control loop:

```
joint_pub.publish(joint_state);
broadcaster.sendTransform(odom_trans);
```

We will add the following to the `arm_state_xacro.launch` file:

```
<launch>
    <arg name="model" />
    <arg name="gui" default="False" />
    <param name="robot_description" command="$(find xacro)/xacro --inorder
$(arg model)" />
    <param name="use_gui" value="$(arg gui)"/>
    <node name="arm_state_publisher_tutorials" pkg="robot_description"
type="arm_state_publisher_tutorials" />
    <!--node name="joint_state_publisher" pkg="joint_state_publisher"
type="joint_state_publisher"/-->
    <node name="robot_state_publisher" pkg="robot_state_publisher"
type="state_publisher" />
    <node name="rviz" pkg="rviz" type="rviz" args="-d $(find
robot_description)/arm.rviz" />
</launch>
```

Similarly, we will add the following to the `mobile_state_xacro.launch` file:

```
<launch>
    <arg name="model" />
    <arg name="gui" default="False" />
    <param name="robot_description" command="$(find xacro)/xacro --inorder
$(arg model)" />
    <param name="use_gui" value="$(arg gui)"/>
    <node name="mobile_state_publisher_tutorials" pkg="robot_description"
type="mobile_state_publisher_tutorials" />
    <!--node name="joint_state_publisher" pkg="joint_state_publisher"
type="joint_state_publisher"/-->
    <node name="robot_state_publisher" pkg="robot_state_publisher"
type="state_publisher" />
    <node name="rviz" pkg="rviz" type="rviz" args="-d $(find
robot_description)/mobile.rviz" />
</launch>
```

Proceeding further, we have to install the following packages:

```
$ sudo apt-get install ros-kinetic-map-server
$ sudo apt-get install ros-kinetic-fake-localization
```

Compile the packages in the workspace using the following code:

```
$ cd ~/catkin_ws
$ catkin_make
```

Hence, using the following command, we will start the 3D model of the mobile robot in `rviz` and can visualize its circular motion:

```
$ roslaunch robot_description mobile_state_xacro.launch model:="'rospack
find robot1_description'/urdf/mobile_robot.xacro"
```

In the following screenshot, we can view the circular motion of the mobile robot:

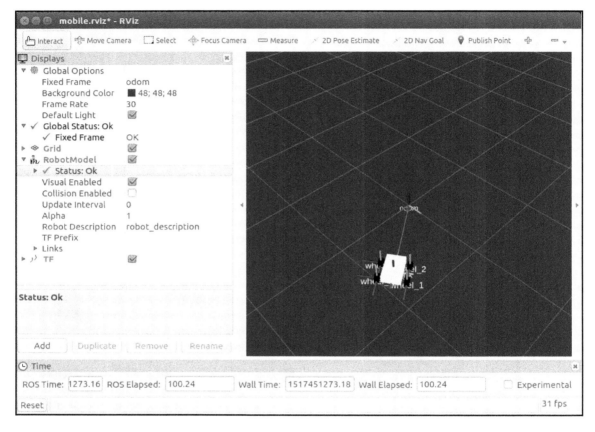

Mobile robot's circular motion

Similarly, using the following command, we will start the 3D model of the arm robot in rviz and can visualize all of its articulations:

```
$ roslaunch robot_description arm_state_xacro.launch model:="'rospack find
robot1_description'/urdf/arm_robot.xacro"
```

In the following screenshot, we can view the articulation of the arm robot:

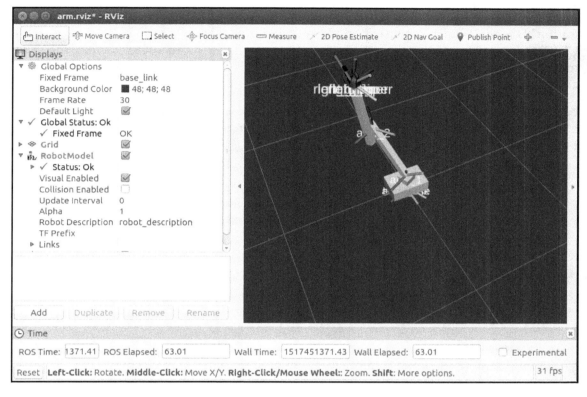

Arm robot articulation

There's more...

3D modeling software such as SketchUp or Blender can be used to generate mesh designs for the 3D model, which can be exported in either `*.dae` or `*.sdf` format. We can create a URDF file and view the model on RViz. However, the detailed process for developing the mesh design for a 3D model using modeling software is beyond the scope of this book, but an advanced reader can find the resources on the web.

Nevertheless, we have the file named `bot.dae` in the `robot_description/meshes` folder at GitHub, which is a basic mesh design of the mobile robot model for demonstration purposes.

Hence, we will learn how to use the mesh design of a 3D model using the URDF file. Therefore, we will create a new file in the `robot_description/urdf` folder with the name `dae.urdf`, and we will add the following code to it (refer to the `robot_description/urdf/dae.urdf` package at GitHub):

```
<?xml version="1.0"?>
<robot name="robot">
  <link name="base_link">
    <visual>
      <geometry>
        <mesh scale="1 1 1"
filename="package://robot_description/meshes/bot.dae"/>
      </geometry>
      <origin xyz="0 0 0.226"/>
    </visual>
  </link>
</robot>
```

The model can be exported to one file only, and so the wheels and chassis base are part of the same object. In order to develop a robot model with mobile parts, each part must be exported to a separate file.

We can verify the model by using the following command:

```
$ roslaunch robot_description simulation.launch model:="'rospack find
robot_description'/urdf/dae.urdf"
```

As a result, we will get the following output:

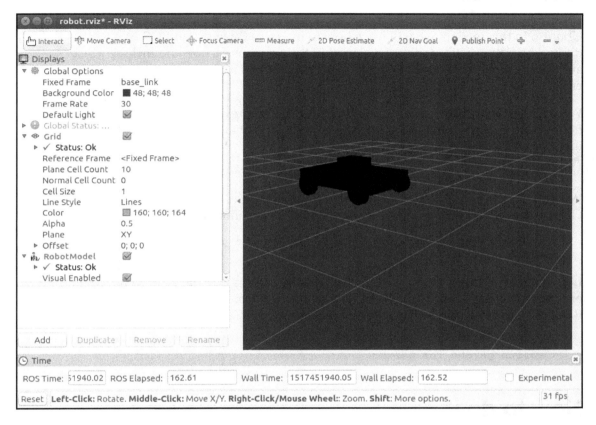

Mesh design

Understanding the Gazebo architecture and interface with ROS

As we discussed in the previous chapter, Gazebo is a simulation framework with a physics engine which is capable of simulating various sets of multi-robots for complex structured and unstructured environments. Moreover, it is capable of simulating several sensors and objects in a three-dimensional world. It also provides both realistic sensor feedback and physical interactions among objects and the environment.

Getting ready

However, Gazebo is independent of ROS and is available as a standalone package for Ubuntu. In this section, we will learn how to interface Gazebo and ROS. We will also discuss how to use 3D models developed in the previous section and include a laser sensor and a camera there. We will also discuss how to control and move the robot in a virtual environment.

We will use the mobile robot model developed in the previous section, although we will not include the arm robot to keep things simple. First of all, we have to confirm the Gazebo installation by executing the following command:

```
$ gazebo
```

We also have to install ROS packages to interface Gazebo, before we start working with Gazebo in the ROS environment, using the following command:

```
$ sudo apt-get install ros-kinetic-gazebo-ros-pkgsros-kinetic-Gazebo-ros-control
```

Let's hope everything works well. Afterward, we can see that the Gazebo GUI opens after executing this command. Moreover, we can also confirm the integration of Gazebo with ROS using the following commands which pop up the Gazebo GUI:

```
$ roscore
$ rosrun gazebo_ros gazebo
```

How to do it...

We will complete the URDF file to introduce the robot model in Gazebo by defining a few more elements there. However, we will be using the .xacro file which is more complex, but more powerful for development. We can find the file with all its modifications at chapter6_tutorials/robot_description/urdf/robot_gazebo.xacro on GitHub (https://github.com/kbipin/Robot-Operating-System-Cookbook):

```
<link name="base_link">
            <visual>
            <geometry>
                    <box size="0.2 .3 .1"/>
                </geometry>
            <origin rpy="0 0 1.54" xyz="0 0 0.05"/>
            <material name="white">
                <color rgba="1 1 1 1"/>
            </material>
```

```
                </visual>
        <collision>
                <geometry>
                        <box size="0.2 .3 0.1"/>
                </geometry>
        </collision>
        <xacro:default_inertial mass="10"/>
</link>
```

You may have noticed in the preceding code that collision and inertial sections are necessary to introduce the robot model to Gazebo, which is required by the physics engine for simulation.

We will also add visible textures in Gazebo by including the `robot.gazebo` file in `robot_gazebo.xacro` as follows:

```
<xacro:include filename="$(find robot_description)/urdf/robot.gazebo" />
```

In the preceding code, `robot.gazebo` has texture information, which is shown as follows; we can refer to the complete file in the `chapter6_tutorials` packages at GitHub (`https://github.com/kbipin/Robot-Operating-System-Cookbook`):

```
<gazebo reference="base_link">
  <material>Gazebo/Orange</material>
</gazebo>

<gazebo reference="wheel_1">
        <material>Gazebo/Black</material>
</gazebo>

<gazebo reference="wheel_2">
        <material>Gazebo/Black</material>
</gazebo>

<gazebo reference="wheel_3">
        <material>Gazebo/Black</material>
</gazebo>

<gazebo reference="wheel_4">
        <material>Gazebo/Black</material>
</gazebo>
```

Then, we will create an integrated `launch` file to launch all components with the name `gazebo.launch` in the `chapter6_tutorials/robot_gazebo/launch/` folder and add the following code (refer to the code in the `chapter6_tutorials` packages at GitHub (`https://github.com/kbipin/Robot-Operating-System-Cookbook`)):

```
<launch>
  <!-- these are the arguments you can pass this launch file, for example
paused:=true -->
  <arg name="paused" default="true" />
  <arg name="use_sim_time" default="false" />
  <arg name="gui" default="true" />
  <arg name="headless" default="false" />
  <arg name="debug" default="true" />
  <!-- We resume the logic in empty_world.launch, changing only the name of
the world to be launched -->
  <include file="$(find gazebo_ros)/launch/empty_world.launch">
    <arg name="world_name" value="$(find robot1_gazebo)/worlds/robot.world"
/>
    <arg name="debug" value="$(arg debug)" />
    <arg name="gui" value="$(arggui)" />
    <arg name="paused" value="$(arg paused)" />
    <arg name="use_sim_time" value="$(arg use_sim_time)" />
    <arg name="headless" value="$(arg headless)" />
  </include>
  <!-- Load the URDF into the ROS Parameter Server -->
  <arg name="model" />
  <param name="robot_description" command="$(find xacro)/xacro.py $(arg
model)" />
  <!-- Run a python script to the send a service call to gazebo_ros to
spawn a URDF robot -->
  <node name="urdf_spawner" pkg="gazebo_ros" type="spawn_model"
respawn="false" output="screen" args="-urdf -model robot1 -
paramrobot_description -z 0.05" />
</launch>
```

We will launch the `robot` model with ROS and Gazebo by using the following command:

```
$ roslaunch robot_gazebo gazebo.launch model:="'rospack find
robot1_description'/urdf/robot_gazebo.xacro"
```

Soon, we will be able to view the robot in Gazebo, as shown in the following screenshot. Here, the simulation is paused initially; we have to start it by clicking the **Play** button, which is at the bottom left of the display bar:

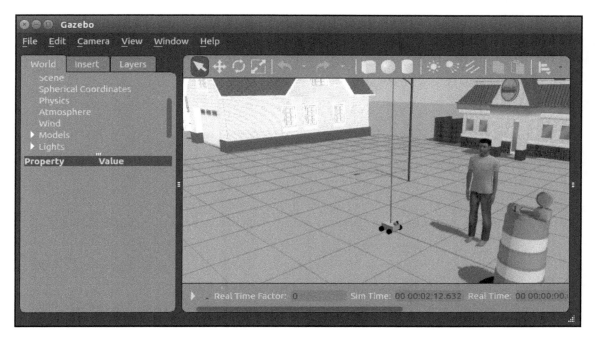

Robot model in the Gazebo virtual world

Welcome! This will be our first step into the virtual world.

Integrating sensors

We have learned how to simulate the physics of the robot and its motion, and we can also simulate the sensors in Gazebo.

Usually, in order to simulate the sensors, we need to implement their physical behavior. In other words, it is required that we design the mathematical model of the sensors.

In this section, we will discuss how to add a camera and a laser sensor to the mobile robot model we designed previously. Since these sensors are new elements of the robot model, first of all, we have to decide where to install them. In the preceding screenshot, you may have noticed a new 3D gadget that looks like a Hokuyo laser, and a red cube which will be the camera in Gazebo.

We will use the laser simulation plugin from the `gazebo_ros_demos` package. Accordingly, we have to include the following lines in the `robot_gazebo.xacro` file to add the 3D model of a Hokuyo laser to the mobile robot:

```
<link name="hokuyo_link">
  <collision>
    <origin xyz="0 0 0" rpy="0 0 0"/>
    <geometry>
  <box size="0.1 0.1 0.1"/>
    </geometry>
  </collision>
  <visual>
    <origin xyz="0 0 0" rpy="0 0 0"/>
    <geometry>
      <mesh filename="package://robot_description/meshes/hokuyo.dae"/>
    </geometry>
  </visual>
  <inertial>
    <mass value="1e-5" />
    <origin xyz="0 0 0" rpy="0 0 0"/>
    <inertia ixx="1e-6" ixy="0" ixz="0" iyy="1e-6" iyz="0" izz="1e-6" />
  </inertial>
</link>
```

Furthermore, we will add the `libgazebo_ros_laser.so` plugin to the `robot.gazebo` file, which will simulate the behavior of a Hokuyo range laser in Gazebo. Similarly, we will add the `libgazebo_ros_camera.so` plugin there to simulate the camera. We can refer to the source code of `robot.gazebo` in the `chapter6_tutorials` packages at GitHub (`https://github.com/kbipin/Robot-Operating-System-Cookbook`):

```
<!-- hokuyo -->
  <gazebo reference="hokuyo_link">
    <sensor type="ray" name="head_hokuyo_sensor">
      <pose>0 0 0 0 0 0</pose>
      <visualize>false</visualize>
      <update_rate>40</update_rate>
      <ray>
        <scan>
          <horizontal>
            <samples>720</samples>
            <resolution>1</resolution>
            <min_angle>-1.570796</min_angle>
            <max_angle>1.570796</max_angle>
          </horizontal>
        </scan>
        <range>
```

```
            <min>0.10</min>
            <max>30.0</max>
            <resolution>0.01</resolution>
          </range>
          <noise>
            <type>gaussian</type>
            <!-- Noise parameters based on published spec for Hokuyo laser
                 achieving "+-30mm" accuracy at range < 10m.  A mean of 0.0m
and
                 stddev of 0.01m will put 99.7% of samples within 0.03m of
the true
                 reading. -->
            <mean>0.0</mean>
            <stddev>0.01</stddev>
          </noise>
        </ray>
        <plugin name="gazebo_ros_head_hokuyo_controller"
filename="libgazebo_ros_laser.so">
          <topicName>/robot/laser/scan</topicName>
          <frameName>hokuyo_link</frameName>
        </plugin>
      </sensor>
    </gazebo>

    <!-- camera -->
    <gazebo reference="camera_link">
      <sensor type="camera" name="camera1">
        <update_rate>30.0</update_rate>
        <camera name="head">
          <horizontal_fov>1.3962634</horizontal_fov>
          <image>
            <width>800</width>
            <height>800</height>
            <format>R8G8B8</format>
          </image>
          <clip>
            <near>0.02</near>
            <far>300</far>
          </clip>
          <noise>
            <type>gaussian</type>
            <!-- Noise is sampled independently per pixel on each frame.
                 That pixel's noise value is added to each of its color
                 channels, which at that point lie in the range [0,1]. -->
            <mean>0.0</mean>
            <stddev>0.007</stddev>
          </noise>
        </camera>
```

```
<plugin name="camera_controller" filename="libgazebo_ros_camera.so">
  <alwaysOn>true</alwaysOn>
  <updateRate>0.0</updateRate>
  <cameraName>robot/camera1</cameraName>
  <imageTopicName>image_raw</imageTopicName>
  <cameraInfoTopicName>camera_info</cameraInfoTopicName>
  <frameName>camera_link</frameName>
  <hackBaseline>0.07</hackBaseline>
  <distortionK1>0.0</distortionK1>
  <distortionK2>0.0</distortionK2>
  <distortionK3>0.0</distortionK3>
  <distortionT1>0.0</distortionT1>
  <distortionT2>0.0</distortionT2>
</plugin>
</sensor>
</gazebo>
```

Finally, we will launch the updated model with the following command:

```
$ roslaunch robot_gazebo gazebo.launch model:="'rospack find
robot1_description'/urdf/robot_gazebo.xacro.xacro"
```

In the following screenshot, we can see the robot model with the simulated Hokuyo laser sensor as a small cylinder in black at the top and a red cube beside it that simulates the camera model:

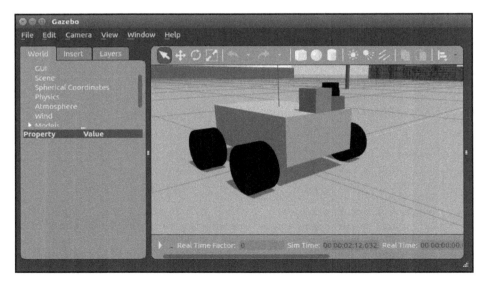

Laser and camera

It is exciting to recognize that these simulated laser and camera sensors are generating real data, which can be viewed by using the `rostopic echo` command as follows:

```
$ rostopic echo /robot/laser/scan
$ rosrun image_view image_view image:=/robot/camera1/image_raw
```

The following screenshot shows the laser and camera output:

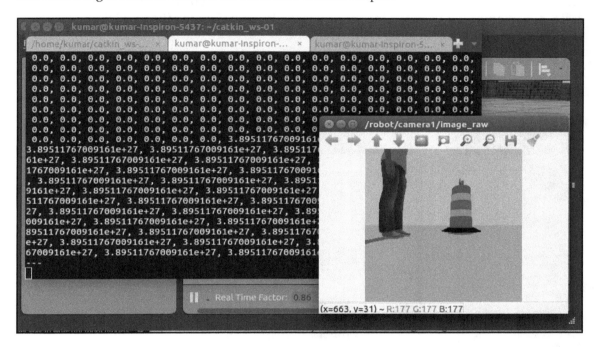

Laser and camera output

Using map

In Gazebo, we can develop our own virtual worlds such as offices, residential space, mountains, and so on, using the SDF file format. Here, we will use a map of a residential space that has been created using the Gazebo GUI and saved as `robot.world` (SDF file format) in the `chapter6_tutorials/robot_gazebo/` folder. The robot with a map of residential space is shown in the preceding screenshot.

There are subtle differences between URDF and SDF file formats, despite both using XML syntax. URDF works with the rest of the ROS, whereas SDF really only works with Gazebo. Additionally, URDF can be converted into SDF specifically for Gazebo. However, SDF is more flexible and powerful with Gazebo for simulating realistic virtual worlds and models.

Additionally, the `gazebo_worlds` package, which is installed by default, has a set of pre-created map of virtual worlds. One of them is the Willow Garage office, which can be launched by using the following command:

```
$ roslaunch gazebo_ros willowgarage_world.launch
```

We can view the expected output, which is shown in the following screenshot:

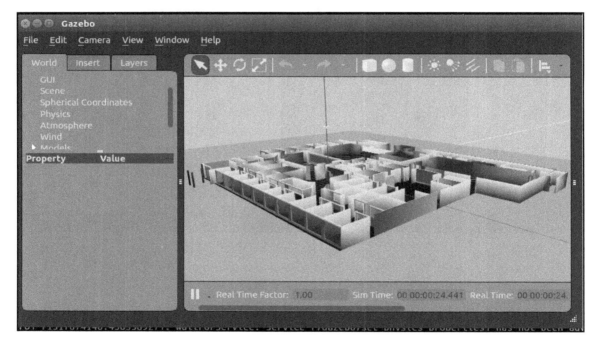

Willow Garage office in simulation

Gazebo requires a good, powerful machine, with a relatively recent GPU. However, sometimes, the simulation crashes during startup, but communities are putting great effort into making it stable with each upcoming release. Usually, if crash problems persist, even for multiple trails, it would be advisable to work with a newer version.

Controlling the robot

The differential drive robot is also known as a skid-steer robot, whose motion is defined by separately driven wheels fixed on either side of the robot's body. There is no steering mechanism, which is its primary design, but we can change its direction by varying the relative rate of rotation of its wheels on either side.

Gazebo requires that we program the behaviors of the robot, joints, and sensors in the virtual world. It also provides a skid drive plugin called `libgazebo_ros_skid_steer_drive.so` to control the differential drive robot, which we designed in the previous section.

We will integrate this controller into the robot model by adding the following lines of code to `robot_gazebo.xacro` (refer to the complete code in GitHub):

```
<!-- Drive controller -->
<gazebo>
  <plugin name="skid_steer_drive_controller"
filename="libgazebo_ros_skid_steer_drive.so">
    <updateRate>100.0</updateRate>
    <robotNamespace>/</robotNamespace>
    <leftFrontJoint>base_to_wheel1</leftFrontJoint>
    <rightFrontJoint>base_to_wheel3</rightFrontJoint>
    <leftRearJoint>base_to_wheel2</leftRearJoint>
    <rightRearJoint>base_to_wheel4</rightRearJoint>
    <wheelSeparation>4</wheelSeparation>
    <wheelDiameter>0.1</wheelDiameter>
    <robotBaseFrame>base_link</robotBaseFrame>
    <torque>1</torque>
    <topicName>cmd_vel</topicName>
    <broadcastTF>0</broadcastTF>
  </plugin>
</gazebo>
```

Here, in the preceding code segment, we will primarily perform the configuration setup for the controller where the `base_to_wheel1`, `base_to_wheel2`, `base_to_wheel3`, and `base_to_wheel4` joints are selected as wheels to move the robot. In addition, we will also configure the `topicName` to publish commands so that we can control the robot. In this case, the `/cmd_vel` topic, which has the `sensor_msgs/Twist` type as the default, is configured to move the mobile robot.

Finally, we will launch the mobile robot model with the controller and the map using the following command:

```
$ roslaunch robot_gazebo gazebo.launch model:="'rospack find
robot1_description'/urdf/robot_gazebo.xacro"
```

We will see the robot with the map on the Gazebo screen, as shown in the preceding screenshot of the previous section. Subsequently, we would like to move the robot using the keyboard. Therefore, we will use the teleop_twist_keyboard package that publishes the /cmd_vel topic upon a key press. The teleop-twist-keyboard package must be installed:

```
$ sudo apt-get install ros-kinetic-teleop-twist-keyboard
```

Then, we will start the node so that it captures the key press and publishes the command velocities:

```
$ rosrun teleop_twist_keyboard teleop_twist_keyboard.py
```

You may have noticed a new terminal with some instructions and the keys to move the robot. The terminal must be in focus or on the top to receive the key inputs:

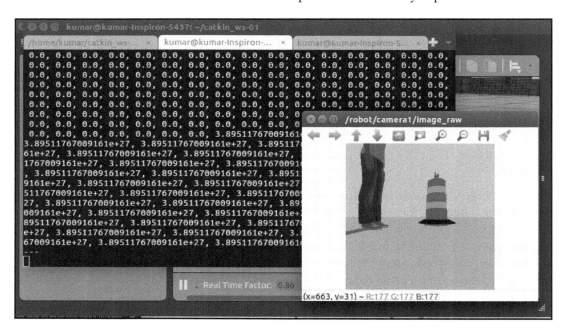

Moving the robot in the virtual world

Amen! Everything has gone well, and we can drive the robot across the residential space. We can also visualize images from the camera and see the laser sensor's output, as shown in the preceding screenshot.

Mobile Robot in ROS 7

In this chapter, we will discuss the following recipes:

- The navigation stack in ROS
- Interfacing the mobile robot to the navigation stack
- Creating a launch file for the navigation stack
- Setting up RViz for the navigation stack – visualization
- Robot localization – Adaptive Monte Carlo Localization (AMCL)
- Configuring navigation stack parameters with rqt_reconfigure
- Autonomous navigation of mobile robots – avoiding obstacles

Introduction

In the previous chapters, we learned how to develop, move, and arm a robot in Gazebo, and mount a few sensors and actuators, as well as move and control them in the virtual world using a joystick or the keyboard. In this chapter, we will discuss one of the most powerful features in the *ROS navigation stack*, which will be equipped with a mobile robot to move autonomously.

First of all, we will learn how to configure the navigation stack with any mobile robot. Subsequently, we will also discuss how to configure and launch the navigation stack on the simulated robot and set configuration parameters to achieve the best results.

Finally, we will look at executing **Simultaneous Localization and Mapping** (**SLAM**) with ROS, which will build the map of the environment while a mobile robot moves through it. In addition, we will also discuss the **Adaptive Monte Carlo Localization** (**AMCL**) algorithm of the navigation stack.

The navigation stack in ROS

The navigation stack in ROS consists of a set of algorithms that uses the sensors and the odometry information from a robot and provides the interface to control the robot using a standard message. The stack can move the robot autonomously in the environment, without crashing, getting stuck in a location, or getting lost elsewhere.

However, it is possible to integrate this stack with any mobile robot, but tuning a few of the configuration parameters is necessary, and this depends upon the robot's specification. Moreover, some of the interface nodes need to be developed so that we can use the stack efficiently.

Getting ready

The mobile robot must satisfy a few requirements before it can be used with the navigation stack:

- The navigation stack can only support a differential drive and holonomic-wheeled robot.
- The shape of the robot must either be a square or a rectangle.
- The robot must provide information about positions of all the joints and sensors and the relationship between their coordinates frames.
- The robot must, at the very least, have a range sensor or similar such as a planar laser or a sonar. However, depth sensors can also be projected as ranger sensors before using them with the stack.

The following diagram shows the architecture of navigation stacks, which have three groups of boxes: gray, white, and dotted lines. The plain white boxes indicate the components of the stack provided by ROS, which has all nodes responsible for autonomous navigation:

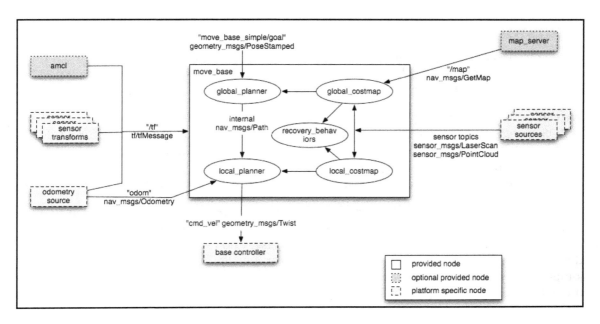

ROS navigation stack

In the following section, we will discuss the development of platform-dependent components of the navigation stack, which are shown in gray boxes.

How it works...

In this section, we will discuss creating transforms between various frames of reference and viewing the transform tree, followed by publishing the sensor and odometry information in Gazebo and a real robot. We will also discuss how to create a base controller to control the mobile robot.

Finally, we will look at working with a map for an environment that has been created by a mobile robot using the ROS navigation stack.

Transform Frames

As we discussed in the previous section, the navigation stack needs information about the position of the sensors, wheels, and joints in the robot body's fixed frame, also known as the `base_link` and relationship between their coordinates frames.

We can recall our learning about the **Transform Frames** (**TF**) software library from Chapter 3, *ROS Architecture and Concepts-II in ROS*, which we can use to manage a transform tree. In other words, we can add more sensors and parts to the robot, and the TF library will handle all the relations for us and perform all the mathematics for us to get a transform tree that defines offsets in terms of both translation and rotation between different reference frames.

We will develop the mobile robot simulation model in Chapter 6, *Robot Modeling and Simulation*, which has camera and laser sensors attached on its chassis. Moreover, the navigation system must know the position of the laser and camera on the robot chassis to detect collisions, such as one between the wheels and walls. Hence, all the sensors and joints must be correctly configured with the TF library to allow the navigation stack to move the robot in a consistent way by knowing exactly where each one of their components resides.

In the case of a real robot, we have to write the code to configure each component and transform it. However, if the URDF of the robot represents the real robot, our simulation will be exactly the same as the real robot, so we do not actually need to configure each component. In our case, for simulating the geometry of the robot specified in the URDF file, it is not necessary to configure the robot again, since we are using the `robot_state_publisher` package to publish the transform tree of the robot.

We could view the transformation tree of the mobile robot developed in the previous chapter by using the following command:

```
$roslaunch chapter7_tutorials gazebo_map_robot.launch model:="'rospack find
chapter7_tutorials'/urdf/robot_model_01.xacro"
$rosrun tf view_frames
```

The following diagram shows the transform tree of our simulated mobile robot:

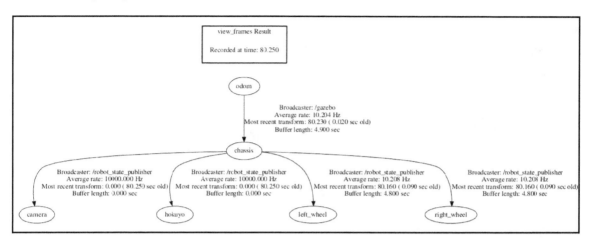

Transform Frame

Sensors

The real robot could have several sensors to perceive the world. We can have many nodes to receive this data and perform processing, whereas the navigation stack can only use the planar range sensor by design. Here, the sensor node must publish the data with one of these types: /sensor_msgs::LaserScan or /sensor_msgs::PointCloud2.

We will use the laser located in front of the simulated mobile robot to navigate the Gazebo world. This laser is simulated on Gazebo, and it publishes data on the hokuyo_link reference frame with the topic name /robot/laser/scan. Here, we do not have to configure anything for the laser to use in the navigation stack since TF already configured it in the .urdf file, and the laser is publishing data the correct way.

In the case of a real laser sensor, we will have to develop a driver for it. Indeed, in Chapter 5, *Accessing Sensors and Actuators through ROS*, we have already discussed how to connect the Hokuyo laser device to ROS.

We can view the workings of the laser sensor in a simulation by using the following command:

```
$roslaunch chapter7_tutorials gazebo_xacro.launch model:="'rospack find
chapter7_tutorials'/urdf/robot_model_04.xacro"
```

The following screenshot shows the laser sensor in Gazebo:

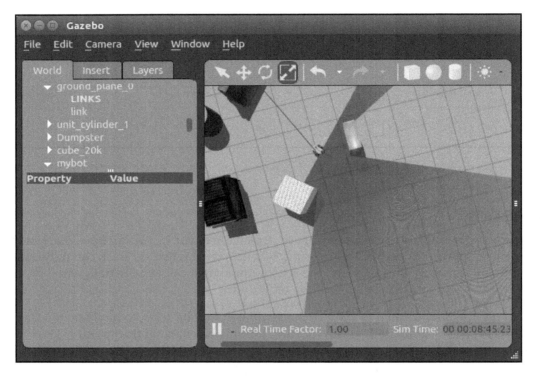

Laser in Gazebo

The following screenshot also shows the visualization of the laser data sensor in RViz:

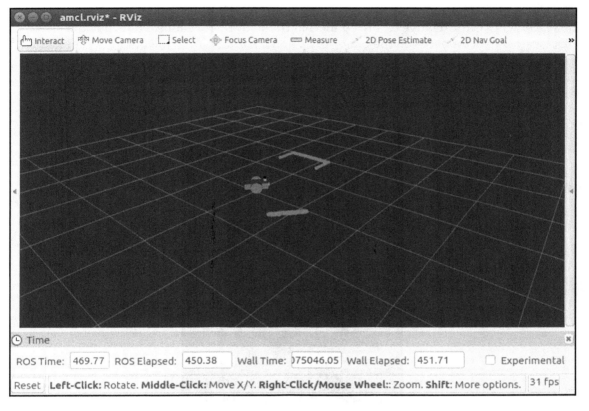

Laser data visualization in RViz

Odometry

The navigation stack requires odometry data from the robot. The odometry is the displacement vector relative to a fixed inertial frame or world frame. Here, it is the displacement vector between the `base_link` and a fixed point in the odom frame. The message type used by the navigation stack is `nav_msgs/Odometry`. The following command shows its detailed structure which is shown in the following screenshot:

```
$ rosmsg show nav_msgs/Odometry
```

```
std_msgs/Header header
  uint32 seq
  time stamp
  string frame_id
string child_frame_id
geometry_msgs/PoseWithCovariance pose
  geometry_msgs/Pose pose
    geometry_msgs/Point position
      float64 x
      float64 y
      float64 z
    geometry_msgs/Quaternion orientation
      float64 x
      float64 y
      float64 z
      float64 w
  float64[36] covariance
geometry_msgs/TwistWithCovariance twist
  geometry_msgs/Twist twist
    geometry_msgs/Vector3 linear
      float64 x
      float64 y
      float64 z
    geometry_msgs/Vector3 angular
      float64 x
      float64 y
      float64 z
  float64[36] covariance
```

Odometry message structure

In the preceding screenshot, we can see that the pose has two structures that represent the position in Euler coordinates and the orientation of the robot using a quaternion. Similarly, the velocity has two structures that represent the linear velocity and the angular velocity. However, the simulated mobile robot we are using has only the linear x velocity and the angular z velocity, since it is simulated as a differential drive model.

As the odometry is defined as the displacement vector between two frames, it requires that it publishes its transform. Since we are working in a virtual world, let's discuss how Gazebo works with odometry.

As we have discussed previously, our robot moves in the simulated world just like a robot in the real world. We configured `diffdrive_plugin` to drive our robot in `Chapter 6`, *Robot Modeling and Simulation*. Since this driver publishes the odometry generated in the simulated world, nothing needs to be developed for Gazebo.

Now, we will start the simulated mobile robot in Gazebo to see how odometry works by running the following command in two separate terminals:

```
$ roslaunch chapter7_tutorials gazebo_xacro.launch model:="'rospack find
robot1_description'/urdf/robot_model_04.xacro"
$ rosrun teleop_twist_keyboard teleop_twist_keyboard.py
```

The output is shown in the following screenshot:

Odometry data in Gazebo

We use the backtick and not a simple ' character for these kinds of instructions. We should be careful here, in case the `'rospack find robot1_description'` instruction does not generate the desired output. We could execute $ `rospack find robot1_description` separately and complete the previous instruction with outputs. For example:

```
$ roslaunch chapter7_tutorials gazebo_xacro.launch
model:=/home/kbipin/catkin_workspace/src/
robot1_description1/urdf/robot_model_04.xacro
```

Then, we would be able to move the robot for a few seconds to generate the new data on the odometry topic using the keyboard.

As shown in the preceding screenshot, on the screen of the Gazebo simulator, we will click the on `robot_model` to view the properties of the object model. Here, one of these properties is the pose of the robot. Similarly, we can click on the pose to view the fields with data, which is the position of the robot in the virtual world. As we move the robot, this data will keep changing, as shown in the preceding screenshot in the yellow bounding box.

Gazebo continuously publishes the odometry data which can be observed by looking at the topic. We can type the following command in a shell to see what data it is sending:

```
$ rostopic echo /odom/pose/pose
```

We will receive the output that is shown in the following screenshot:

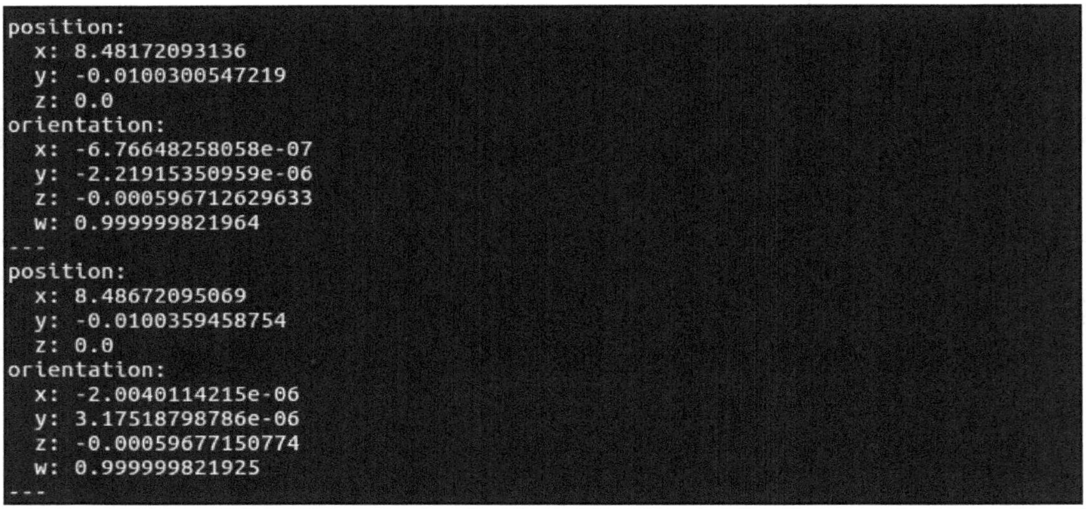

```
position:
  x: 8.48172093136
  y: -0.0100300547219
  z: 0.0
orientation:
  x: -6.76648258058e-07
  y: -2.21915350959e-06
  z: -0.000596712629633
  w: 0.999999821964
---
position:
  x: 8.48672095069
  y: -0.0100359458754
  z: 0.0
orientation:
  x: -2.0040114215e-06
  y: 3.17518798786e-06
  z: -0.00059677150774
  w: 0.999999821925
---
```

Odometry data

We can refer to the code at (`https://github.com/ros-simulation/gazebo_ros_pkgs/blob/kinetic-devel/gazebo_plugins/src/gazebo_ros_skid_steer_drive.cpp`) for more information. We can find out how Gazebo generates the odometry data by looking at the `publishOdometry(double step_time)` function and more.

Once we have learned how and where Gazebo creates the odometry, we will be in a position to develop the code to publish the odometry and the transform for a real robot. However, we will not discuss the real robot here, because this is more platform-dependent.

Base controller

One of the crucial components of the navigation stack is a base controller. It is the only way to effectively control a robot that communicates directly with the hardware of the robot. However, ROS does not provide any common base controller, and so we have to develop a base controller for our mobile robot platform.

The base controller must subscribe to a topic with the name `/cmd_vel`, which has the message type `/geometry_msgs::Twist`. This message can also be used on the odometry message that we saw before. As well as this, the base controller has to generate the correct commands for the robot platform with the correct linear and angular velocities.

We can recall the structure of this message by typing the following command in a shell to view the structure:

```
$ rosmsg show geometry_msgs/Twist
```

The output of the preceding command will be as follows:

```
geometry_msgs/Vector3 linear
    float64 x
    float64 y
    float64 z
geometry_msgs/Vector3 angular
    float64 x
    float64 y
    float64 z
```

Command velocity

In the preceding screenshot, two vector structures show the linear and angular velocities the along x, y, and z axes, respectively, since our mobile robot is based on a differential-wheeled platform that has two motors to move the robot forward and backward and to turn. Therefore, we will only use the linear velocity x and the angular velocity z to drive the robot.

Likewise, we are working with a simulated mobile robot in Gazebo where the base controller is implemented on the plugin driver to move the platform. This means that we won't have to create the base controller for this robot.

Next, let's run our simulated robot on Gazebo to see how the base controller functions. We will have to run the following commands on different terminals:

```
$ roslaunch chapter7_tutorials gazebo_xacro.launch model:="'rospack find chapter7_tutorials'/urdf/robot_model_05.xacro"
$ rosrun teleop_twist_keyboard teleop_twist_keyboard.py
```

When all the nodes are launched and running, we will open rxgraph to view the relation among all the nodes:

```
$ rqt_graph
```

The output is shown in the following screenshot:

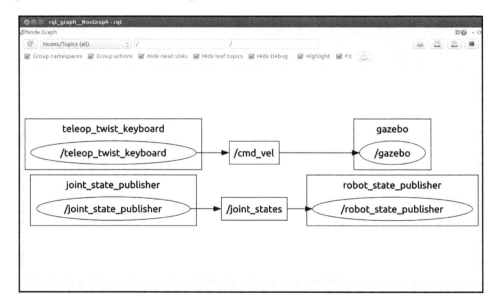

Base controller

In the preceding screenshot, we can see that Gazebo automatically subscribes to the cmd_vel topic that is generated by the teleop node.

As we have discussed in the previous section, the driver plugin of our robot is running inside the Gazebo simulator, which receives data from the cmd_vel topic and moves the robot in the virtual world. Simultaneously, this plugin also generates the odometry data.

Moreover, it can be assumed that we have the adequate background knowledge to develop the base controller for a physical robot. We can also refer to the web for more information (https://www.robotshop.com/).

Map

In this section, we will learn how to use the simulated mobile robot to create a map of the virtual world in Gazebo, save it, and load it again. However, getting a useful map can be a complicated task if we are not using appropriate tools. After all, ROS has a tool named map_server (http://wiki.ros.org/map_server) that can help us build a map of the environment by using odometry and a planer range sensor such as a sensor.

First of all, we will create a .launch file in chapter7_tutorials/launch with the name gazebo_mapping_robot.launch and add the following code:

```xml
<?xml version="1.0"?>
<launch>
<!-- this launch file corresponds to robot model in ros-
pkg/robot_descriptions/pr2/erratic_defs/robots for full erratic -->
<param name="/use_sim_time" value="true" />
<!-- start up wg world -->
<include file="$(find gazebo_ros)/launch/willowgarage_world.launch"/>
<arg name="model" />
<param name="robot_description" command="$(find xacro)/xacro.py $(arg
model)" />
<node name="joint_state_publisher" pkg="joint_state_publisher"
type="joint_state_publisher" ></node>
<!-- start robot state publisher -->
<node pkg="robot_state_publisher" type="robot_state_publisher"
name="robot_state_publisher" output="screen" >
<param name="publish_frequency" type="double" value="50.0" />
</node>
<node name="spawn_robot" pkg="gazebo_ros" type="spawn_model" args="-urdf -
param robot_description -z 0.1 -model robot_model" respawn="false"
output="screen" />
<node name="rviz" pkg="rviz" type="rviz" args="-d $(find
chapter7_tutorials)/launch/mapping.rviz"/>
```

```
<node name="slam_gmapping" pkg="gmapping" type="slam_gmapping">
<remap from="scan" to="/robot/laser/scan"/>
<param name="base_link" value="base_footprint"/>
</node>
</launch>
```

Using this `.launch` file, we can launch the 3D robot model in Gazebo, the RViz program with the appropriate configuration file, and `slam_mapping` to build a map in real time. Subsequently, we will execute this launch in one terminal and run the `teleop` node to move the robot in another terminal as follows:

```
$ roslaunch chapter7_tutorials gazebo_mapping_robot.launch
model:="'rospack find chapter7_tutorials'/urdf/robot1_base_04.xacro"
$ rosrun teleop_twist_keyboard teleop_twist_keyboard.py
```

As shown in the following screenshot, while moving the robot using the keyboard, we can see the free and unknown space on the RViz screen, as well as the map with the occupied space that is known as an **Occupancy Grid Map** (**OGM**). Correspondingly, the `slam_mapping` node updates the map as the robot moves and perceives new information from its surroundings. The `slam_mapping` node needs a good estimate of the robot's location before building the map, which takes the laser scans and the odometry to build the OGM:

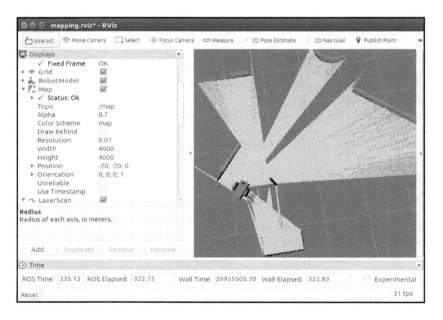

Occupancy Grid Map

Once we have completed building the map so that it is reasonably acceptable, we will save it for later use in autonomous navigation. The following command will save the currently built map:

```
$ rosrun map_server map_saver -f map
```

The preceding command will also create two files, map.pgm and map.yaml. The first one is the map in the .pgm format while the other is the configuration file for the map. The following screenshot shows the contents of map.yaml:

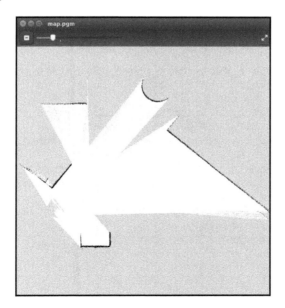

map.yaml

Similarly, we can view the map.pgm file in any of our favorite image viewers, which is shown in the following screenshot:

map.pgm

Whenever we want to use a map that has been built with our robot for a navigation task, we need to load it with the `map_server` packages. The following command will load the map:

```
$ rosrun map_server map_server map.yaml
```

In addition, we would create another `.launch` file in `chapter7_tutorials/launch` with the name `gazebo_map_robot.launch`, and add the following code:

```xml
<?xml version="1.0"?>
<launch>
<!-- this launch file corresponds to robot model in ros-
pkg/robot_descriptions/pr2/erratic_defs/robots for full erratic -->
<arg name="paused" default="true"/>
<arg name="use_sim_time" default="false"/>
<arg name="gui" default="true"/>
<arg name="headless" default="false"/>
<arg name="debug" default="false"/>
<!-- start up wg world -->
<include file="$(find gazebo_ros)/launch/empty_world.launch">
<arg name="debug" value="$(arg debug)" />
<arg name="gui" value="$(arg gui)" />
<arg name="paused" value="$(arg paused)"/>
<arg name="use_sim_time" value="$(arg use_sim_time)"/>
<arg name="headless" value="$(arg headless)"/>
</include>
<arg name="model" />
<param name="robot_description" command="$(find xacro)/xacro.py $(arg
model)" />
<node name="joint_state_publisher" pkg="joint_state_publisher"
type="joint_state_publisher" ></node>
<!-- start robot state publisher -->
<node pkg="robot_state_publisher" type="robot_state_publisher"
name="robot_state_publisher" output="screen" >
<param name="publish_frequency" type="double" value="50.0" />
</node>
<node name="spawn_model" pkg="gazebo_ros" type="spawn_model" args="-urdf -
param robot_description -z 0.1 -model robot_model" respawn="false"
output="screen" />
<node name="map_server" pkg="map_server" type="map_server" args=" $(find
chapter7_tutorials)/maps/map.yaml" />
<node name="rviz" pkg="rviz" type="rviz" />
</launch>
```

Now, we can launch the file using the following command with the correct robot model:

```
$ roslaunch chapter7_tutorials gazebo_map_robot.launch model:="'rospack
find chapter7_tutorials'/urdf/robot_model_04.xacro"
```

At the moment, we can view the robot model with the map in RViz. The navigation stack will use the map published by the map server and the laser reading to perform the localization of the robot using a scan matching algorithm known as AMCL.

We will discuss more about the maps and robot localization in the coming section.

Interfacing the mobile robot to the navigation stack

In the previous section, we learned how to configure our robot so that it can be used with the navigation stack, and in this section, we will learn how to configure the navigation stack and its integration with the robot. All the work done in the previous section has been a prerequisite to this. This is the time the robot comes alive.

Getting ready

In this section, we will discuss and learn the following:

- Understand how the navigation stack works
- Configure all the necessary parameters of the navigation stack using configuration files
- Develop launch files to start the navigation stack with our robot

Before we begin, download the `chapter7_tutorial` source code from GitHub (`https://github.com/kbipin/Robot-Operating-System-Cookbook/tree/master/chapter7_tutorials`).

Alternatively, we can also create them in our workspace.

How to do it...

1. Begin by creating a new file in `chapter7_tutorials/launch` with the name `chapter7_configuration_gazebo.launch`, and add the following code:

```
<?xml version="1.0"?>
<launch>
<param name="/use_sim_time" value="true" />
<remap from="robot/laser/scan" to="/scan" />
<!-- start up wg world -->
<include file="$(find gazebo_ros)/launch/willowgarage_world.launch">
</include>
<arg name="model" default="$(find
chapter7_tutorials)/urdf/robot_model_05.xacro"/>
<param name="robot_description" command="$(find xacro)/xacro.py $(arg
model)" />
<node name="joint_state_publisher" pkg="joint_state_publisher"
type="joint_state_publisher" ></node>
<!-- start robot state publisher -->
<node pkg="robot_state_publisher" type="robot_state_publisher"
name="robot_state_publisher" output="screen" />
<node name="spawn_robot" pkg="gazebo_ros" type="spawn_model" args="-urdf -
param robot_description -z 0.1 -model robot_model" respawn="false"
output="screen" />
<node name="rviz" pkg="rviz" type="rviz" args="-d $(find
chapter7_tutorials)/launch/navigation.rviz" />
</launch>
```

This `launch` file performs the robot's configuration, which is required regarding integration with the navigation stack. Moreover, this launch file is very similar to what we discussed and used in the previous section.

2. Execute the launch file as follows:

```
$ roslaunch chapter7_tutorials chapter7_configuration_gazebo.launch
```

We will get the following window appear:

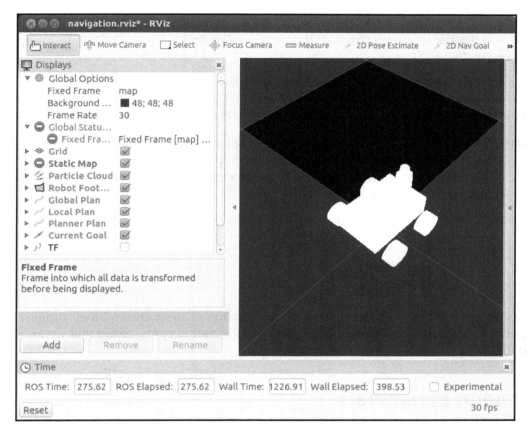

Navigation stack configuration

In the preceding screenshot, there are some fields in red, blue, and yellow, since we have not configured the navigation stack yet. Nevertheless, in coming sections, we will discuss how to configure `rviz` so that we can use it with the navigation stack and view all the topics.

How it works...

Now, we will learn how to configure the navigation stack and all the necessary files. First of all, we will discuss costmap, and how the mobile robot moves through the map using two types of navigation strategy: global and local.

The global navigation is used to create paths from the start to a goal location in the map or at a far-off distance, whereas local navigation is used to create paths in the nearby distance to avoid obstacles, for example, with a square window of 4 x 4 meters around the robot.

Similarly, the `globalcostmap` is used for global navigation and the `localcostmap` is used for local navigation. The costmaps have parameters to configure that define the behaviors for the robot during navigation. Some of the parameters are common to both global and local costmaps, which can be defined in the share file.

Moreover, the configuration mostly consists of three files, as follows:

* `costmap_common_params.yaml`
* `global_costmap_params.yaml`
* `local_costmap_params.yaml`

At this moment in time, we have a basic understanding of costmaps and their usages. We are going to create the configuration files and discuss various parameters that we are going to configure.

Common parameters

Let's begin with the common parameters. First of all, we will create a file with the name `costmap_common_params.yaml` in `chapter6_tutorials/launch` and add the following code (refer to GitHub for more information `https://github.com/kbipin/Robot-Operating-System-Cookbook/blob/master/chapter7_tutorials/launch/costmap_common_params.yaml`):

```
obstacle_range: 2.5
raytrace_range: 3.0
footprint: [[-0.2,-0.2],[-0.2,0.2], [0.2, 0.2], [0.2,-0.2]]
#robot_radius: ir_of_robot
inflation_radius: 0.5
cost_scaling_factor: 10.0
observation_sources: scan
scan: {sensor_frame: base_link, observation_persistence: 0.0,
max_obstacle_height: 0.4, min_obstacle_height: 0.0, data_type: LaserScan,
topic: /scan, marking: true, clearing: true}
```

The preceding file is used to configure the common parameters that are used in both `local_costmap` and `global_costmap`.

The `obstacle_range` and `raytrace_range` attributes are used to define the maximum distance reading for a sensor that can add new information in the costmaps. The `footprint` attribute is used to define the geometry of the robot for the navigation stack. Moreover, the `cost_scaling_factor` attribute defines the behavior of the robot around the obstacles, aggressive or conservative.

Similarly, the `observation_sources` attribute is used to set the sensors used by the navigation stack to perceive the real world.

However, we are using a simulated laser in Gazebo in our case, but we can also use a point cloud here:

```
{sensor_frame: base_link, observation_persistence: 0.0,
max_obstacle_height: 0.4, min_obstacle_height: 0.0, data_type: LaserScan,
topic: /scan, marking: true, clearing: true}
```

Where the laser is configured to add and clear obstacles in the costmap. We can also add a sensor with a wide range to find obstacles and another sensor to navigate and clear the obstacles. We can refer to the ROS navigation stack tutorials for more information.

Global costmap

Next, we ware going to configure the global costmap configuration file. Here, we will create a new file in `chapter7_tutorials/launch` with the name `global_costmap_params.yaml`, and add the following code:

```
global_costmap:
global_frame: /map
robot_base_frame: /base_footprint
update_frequency: 1.0
static_map: true
```

The `global_frame` and the `robot_base_frame` attributes define the transformation between the map and the robot that is used for the global costmap. We can also configure the update frequency for the costmap, which is 1 Hz here.

Similarly, the `static_map` attribute is used for the global costmap to define whether a map or the map server is used to initialize the costmap. If we are not using a static map, this parameter must be set to false.

Local costmap

After configuring the global costmap, we are going to configure the local costmap. Thus, we will create a new file in `chapter7_tutorials/launch` with the name `local_costmap_params.yaml`, and add the following code:

```
local_costmap:
global_frame: /map
robot_base_frame: /base_footprint
update_frequency: 5.0
publish_frequency: 2.0
static_map: false
rolling_window: true
width: 5.0
height: 5.0
resolution: 0.02
tranform_tolerance: 0.5
planner_frequency: 1.0
planner_patiente: 5.0
```

The `global_frame`, `robot_base_frame`, `update_frequency`, and `static_map` parameters are the same as described during configuring the global costmap, in the previous section. In addition, the `publish_frequency` parameter defines updating the frequency and the `rolling_window` parameter describes that the costmap will be centered on the robot during navigation.

Similarly, the `transform_tolerance` parameter configures the maximum latency for the transforms and the `planner_frequency` parameter configures the planning loop rate. The `planner_patiente` parameter configures the waiting time for the planner to find a valid plan before space-clearing operations are performed.

Finally, we can configure the dimensions and the resolution of the costmap with the `width`, `height`, and `resolution` parameters, which are in meters.

Configuring the planner

Once we have configured the global and local costmap, we need to configure the base planner. This is used to generate the velocity commands to control the mobile robot. Hence, we will create a new file in `chapter7_tutorials/launch` with the name `base_local_planner_params.yaml`, and add the following code:

```
TrajectoryPlannerROS:
max_vel_x: 0.2
```

```
min_vel_x: 0.05
max_rotational_vel: 0.15
min_in_place_rotational_vel: 0.01
min_in_place_vel_theta: 0.01
max_vel_theta: 0.15
min_vel_theta: -0.15
acc_lim_th: 3.2
acc_lim_x: 2.5
acc_lim_y: 2.5
holonomic_robot: false
```

Here, the `config` file defines the maximum and minimum velocities, as well as the acceleration for the mobile robot. The `holonomic_robot` parameter is set to true only if we are using a holonomic platform for the mobile robot. We are working with differential drive platforms, which are non-holonomic.

A holonomic vehicle is one that can move in all the configured space from any position.

Creating a launch file for the navigation stack

In the previous section, we discussed how to configure the navigation stack. We now have to create a `launch` file to start the navigation stack with all its configuration.

Getting ready

We will create a new file in the `chapter7_tutorials/launch` folder with the name `move_base.launch` and put the following code there:

```
<?xml version="1.0"?>
<launch>
<!-- Run the map server -->
<node name="map_server" pkg="map_server" type="map_server" args="$(find
chapter7_tutorials)/maps/map.yaml" output="screen"/>
<include file="$(find amcl)/examples/amcl_diff.launch" >
</include>
<node pkg="move_base" type="move_base" respawn="false" name="move_base"
output="screen">
```

```
<param name="controller_frequency" value="10.0"/>
<param name="controller_patiente" value="15.0"/>
<rosparam file="$(find
chapter7_tutorials)/launch/costmap_common_params.yaml" command="load"
ns="global_costmap" />
<rosparam file="$(find
chapter7_tutorials)/launch/costmap_common_params.yaml" command="load"
ns="local_costmap" />
<rosparam file="$(find
chapter7_tutorials)/launch/local_costmap_params.yaml" command="load" />
<rosparam file="$(find
chapter7_tutorials)/launch/global_costmap_params.yaml" command="load" />
<rosparam file="$(find
chapter7_tutorials)/launch/base_local_planner_params.yaml" command="load"
/>
</node>
</launch>
```

You may have noticed that in this launch file, we have included all configuration files that were created in the previous section. Additionally, we will also launch a map server with a map, which was created in the previous *The navigation stack in ROS* and AMCL node section.

Since our robot has a differential drive model, the amcl node used here is for differential drive robots. If we have holonomic robots, we will need to use the `amcl_omni.launch` file.

How it works...

Before launching this integrated launch file, we must launch the `chapter7_configuration_gazebo launch` file. We will run the following commands in two separate terminals:

```
$ roslaunch chapter7_tutorials chapter7_configuration_gazebo.launch
$ roslaunch chapter7_tutorials move_base.launch
```

Congratulations! We will see the following window:

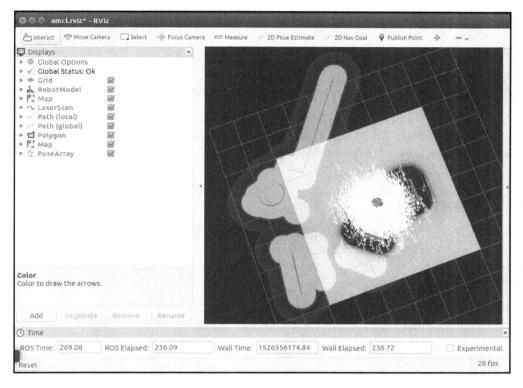

Navigation stack with the robot

You will have noticed that all the options in RViz are in blue; this is a good sign and it means that everything is fine, well done!

In the next section, we will discuss various options used in RViz to visualize all the topics that have been used in the navigation stack in real-time.

Setting up RViz for the navigation stack – visualization

It will always be good practice to visualize all possible data and topics that the navigation stack has used. In this section, we will discuss configuring RViz to visualize the data sent by each of these topics and examine their correctness.

Getting ready

We will run the following commands in two separate terminals to start the navigation stack with our simulated mobile robot in Gazebo, with the map of the virtual environment and the RViz tool for visualization:

```
$ roslaunch chapter7_tutorials chapter7_configuration_gazebo.launch
$ roslaunch chapter7_tutorials move_base.launch
```

As seen in the previous section, we will get the same window that was shown in the preceding screenshot for RViz, and we will have the Gazebo virtual world and simulated mobile robot in another window.

How it works...

We are going to discuss how to configure each of the following topics in RViz, which are crucial for autonomous mobile robot running with the ROS navigation stack. The autonomous mobile robot can drive to the goal and avoid obstacles autonomously in an unstructured environment.

2D pose estimate

The 2D pose estimate allows the user to initialize the localization system of the navigation stack by setting the pose of the robot in the world. The navigation stack waits on the topic with the name /initialpose for the first pose, which can be sent by using RViz windows with the click of a mouse. If this wasn't done at the beginning, the robot will start the auto-localization process and try to set an initial pose.

We can look upon the following screenshot for setting the initial pose. Here we will click on the **2D Pose Estimate** button, and click on the map to point to the initial position of our robot:

- **Topic**: initialpose
- **Type**: geometry_msgs/PoseWithCovarianceStamped

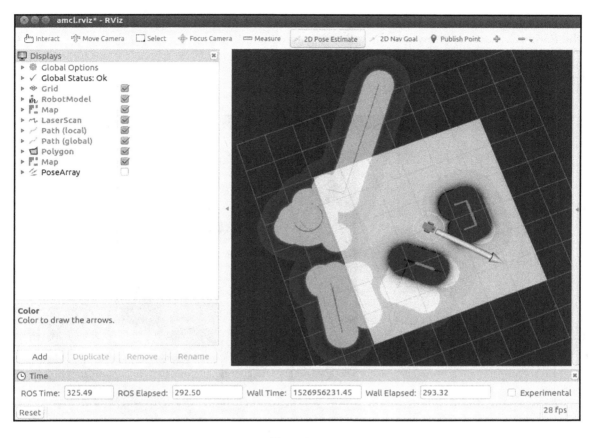

2D pose estimate

2D nav goal

The **2D nav goal** (*G* shortcut) allows the user to send a goal to the navigation system. The navigation stack waits on the topic with the name /move_base_simple/goal for a new goal which can be sent by using RViz windows.

We can click on the **2D Nav Goal** button and select the map and the goal for our robot in the RViz window, as shown in the following screenshot:

- **Topic:** `move_base_simple/goal`
- **Type:** `geometry_msgs/PoseStamped`

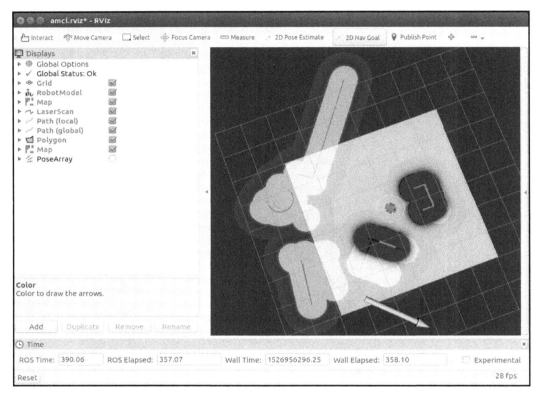

2D nav goal

Static map

By using the `map_server` node that we started in the `launch` file, we will send the static map to RViz for visualization. We can view the map created in the previous section in the RViz window, which is shown in the preceding screenshot.

- **Topic:** `map`
- **Type:** `nav_msgs/GetMap`

Particle cloud

This displays the particle cloud used by the robot's localization system where the of the cloud represents the localization system's uncertainty about the robot's pose. We will obtain the following cloud for our robot, which is shown in the following screenshot:

- **Topic**: `particlecloud`
- **Type**: `geometry_msgs/PoseArray`

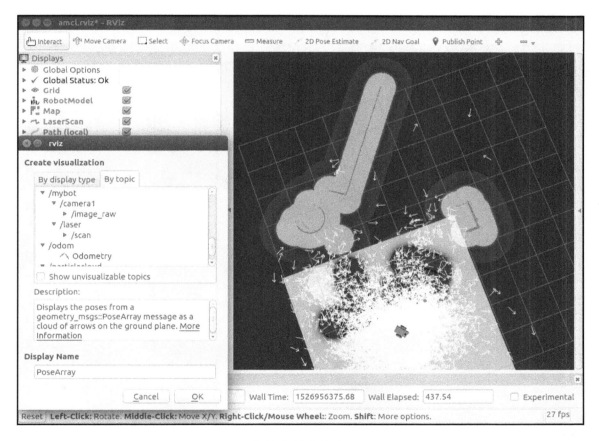

Particle cloud for localization

Robot's footprint

This shows the footprint of the robot where the `width` and `height` parameters of the robot are configured in the `costmap_common_params` file. Thus, this dimension is important because the navigation stack can use it to avoid collisions:

- **Topic**: `local_costmap/robot_footprint`
- **Type**: `geometry_msgs/Polygon`

The following screenshot shows the footprint of our robot:

Robot footprint

Local costmap

This displays the local costmap with the navigation stack that is used for navigation, where the pink line shows the detected obstacle and the blue zone represents the inflated obstacle, as shown in the following screenshot. To have collision-free navigation, the center point of the robot should never overlap with a grid cell that contains an inflated obstacle:

- **Topic**: `move_base/local_costmap/costmap`
- **Type**: `nav_msgs/OccupancyGrid`

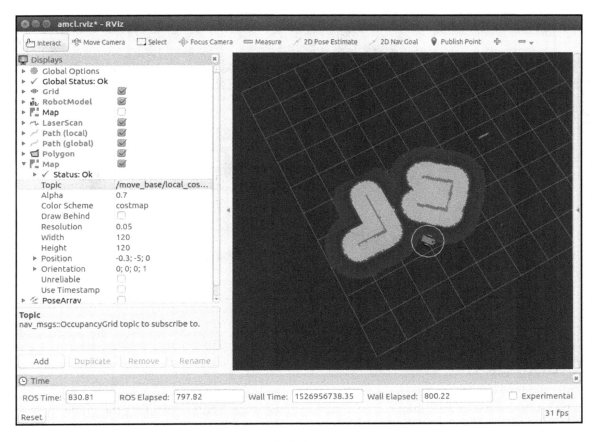

Local costmap

Global costmap

Likewise, this shows the global costmap that the navigation stack uses for navigation, where the pink line shows the detected obstacle and the blue zone represents the inflated obstacle, as shown in the following screenshot. To find the collision-free path, the center point of the robot should never overlap with a cell that contains an inflated obstacle:

- **Topic**: `move_base/global_costmap/costmap`
- **Type**: `nav_msgs/OccupancyGrid`

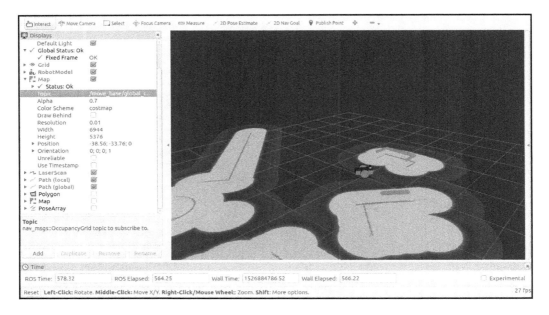

Global costmap

Global plan

This displays the portion of the global plan that the local planner is currently executing, which is shown as a blue line in the following screenshot. While executing the global path, the robot might see obstacles, thereupon the navigation stack will recalculate a new path to avoid collisions, although trying to follow the global plan:

- **Topic**: `TrajectoryPlannerROS/global_plan`
- **Type**: `nav_msgs/Path`

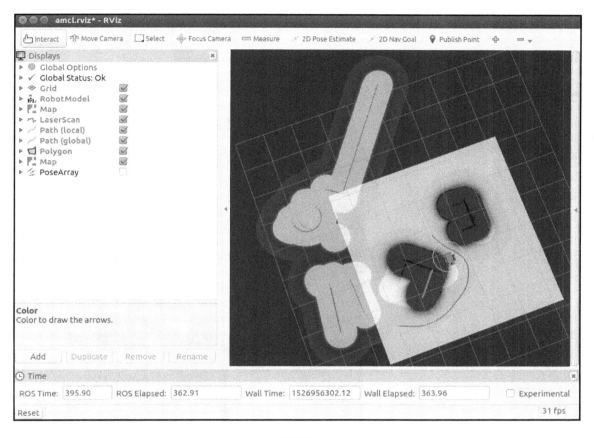

Global plan

Local plan

The local plan shows the trajectory associated with the velocity control commands currently being sent to the base controller by the local planner:

- **Topic**: TrajectoryPlannerROS/local_plan
- **Type**: nav_msgs/Path

We can see that the trajectory in green is in front of the robot, as shown in the following screenshot, which shows whether the robot is moving and the length of the blue line approximate to velocity:

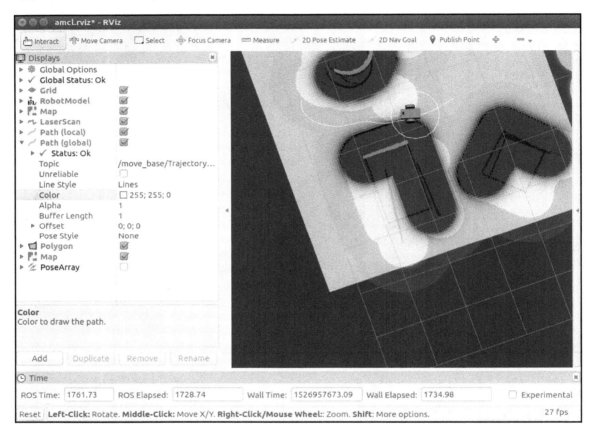

Local plan

Planner plan

The planner plan displays the full plan for the robot, computed by the global planner, which is shown in the following screenshot. It looks very similar to the global plan path:

- **Topic**: NavfnROS/plan
- **Type**: nav_msgs/Path

See the following screenshot:

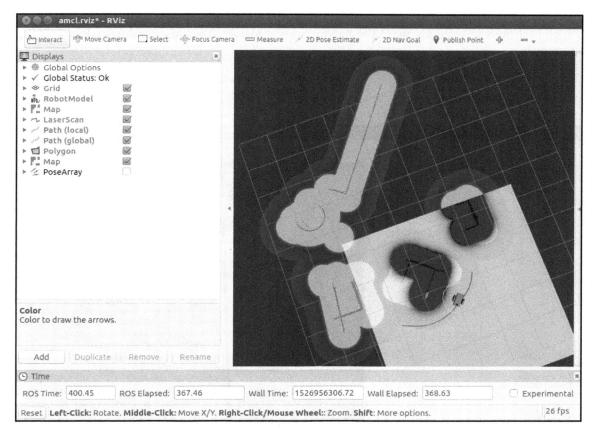

Planner path

Current goal

The current goal shows the goal pose that the navigation stack is attempting to achieve, which is shown as a red arrow in the following screenshot. It represents the final pose of the robot:

- **Topic**: `current_goal`
- **Type**: `geometry_msgs/PoseStamped`

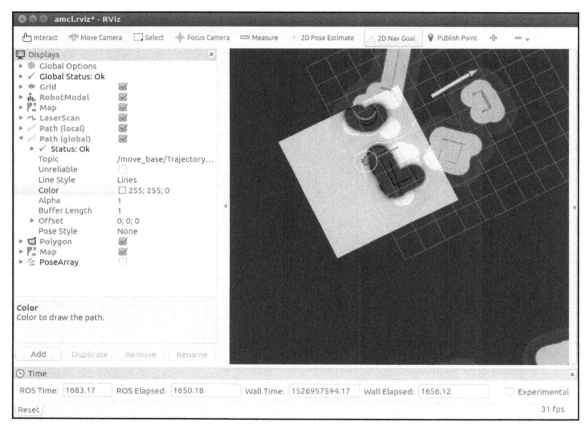

Current goal pose

There's more…

In the previous section, we discussed almost all the visualizations that are necessary to view the navigation stack in RViz. These will be useful to find out whether the robot is doing something unusual. In addition, we can also look for a general ROS navigation system view of the running system using `rqt_graph`, which is shown in the following flow chart:

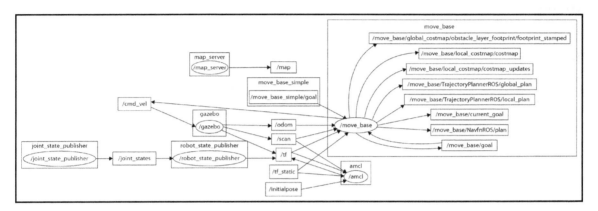

ROS navigation system view

The ROS navigation system view, which is shown in the preceding flow chart, is generated by using the `rqt_graph` tool. This may be not clear in the print version of this book; however, it is placed here for completeness and understanding.

Robot localization – Adaptive Monte Carlo Localization (AMCL)

We are using the AMCL algorithm for robot localization in the given map. The AMCL algorithm is a probabilistic localization technique that uses a particle filter to track the pose of a robot in a known map. It has many configuration options that will affect the performance of localization. We can refer to the AMCL documentation at http://wiki. ros.org/amcl and http://www.probabilistic-robotics.org/.

Getting ready

Although the `amcl` algorithm works mostly with laser scans and laser maps, it can be extended to work with other sensor data, such as stereo vision and lidar, which produce point cloud data. During startup, the `amcl` node initializes its particle filter according to the parameters provided in the setup configuration file. If the initial position of the robot is not set, `amcl` will start at the origin of the reference frame. However, we can set the initial position in RViz using the **2D Pose Estimate** button, as discussed in the previous section.

How it works...

Here, we are using the `amcl_diff.launch` file since our robot has a differential drive platform or model. Moreover, we will discuss some of the parameters in the `amcl_diff.launch` file:

```
<launch>
<node pkg="amcl" type="amcl" name="amcl" output="screen">
<!-- Publish scans from best pose at a max of 10 Hz -->
<param name="odom_model_type" value="diff" />
<param name="odom_alpha5" value="0.1" />
<param name="transform_tolerance" value="0.2" />
<param name="gui_publish_rate" value="10.0" />
<param name="laser_max_beams" value="30" />
<param name="min_particles" value="500" />
<param name="max_particles" value="5000" />
<param name="kld_err" value="0.05" />
<param name="kld_z" value="0.99" />
<param name="odom_alpha1" value="0.2" />
<param name="odom_alpha2" value="0.2" />
<!-- translation std dev, m -->
<param name="odom_alpha3" value="0.8" />
<param name="odom_alpha4" value="0.2" />
<param name="laser_z_hit" value="0.5" />
<param name="laser_z_short" value="0.05" />
<param name="laser_z_max" value="0.05" />
<param name="laser_z_rand" value="0.5" />
<param name="laser_sigma_hit" value="0.2" />
<param name="laser_lambda_short" value="0.1" />
<param name="laser_lambda_short" value="0.1" />
<param name="laser_model_type" value="likelihood_field" />
<!--<param name="laser_model_type" value="beam"/> -->
<param name="laser_likelihood_max_dist" value="2.0" />
<param name="update_min_d" value="0.2" />
<param name="update_min_a" value="0.5" />
```

```
<param name="odom_frame_id" value="odom" />
<param name="resample_interval" value="1" />
<param name="transform_tolerance" value="0.1" />
<param name="recovery_alpha_slow" value="0.0" />
<param name="recovery_alpha_fast" value="0.0" />
</node>
</launch>
```

Here, the `min_particles` and `max_particles` parameters set the minimum and the maximum number of particles allowed for the algorithm. With more particles, accuracy will increase, but this will also increase CPU usage.

The `laser_model_type` parameter configures the laser type. We are using the `likelihood_field` parameter here, but the algorithm can also use beam lasers.

Interestingly, the `initial_pose_x`, `initial_pose_y`, and `initial_pose_a` parameters are not in the launch file, although they set the initial position of the robot when `amcl` starts. For example, if we want, our robot could always start in the dock station; we have to set this position in the launch file.

In addition, we can refer to http://wiki.ros.org/amcl, which has a lot of information about the configuration and parameters for amcl.

Configuring navigation stack parameters with rqt_reconfigure

ROS provides the `rqt_reconfigure` tool for viewing all parameters and configuring them in real-time without restarting the simulation.

How it works...

We will launch `rqt_reconfigure` by using the following command:

```
$ rosrun rqt_reconfigure rqt_reconfigure
```

We will get the same output as that shown in the following screenshot:

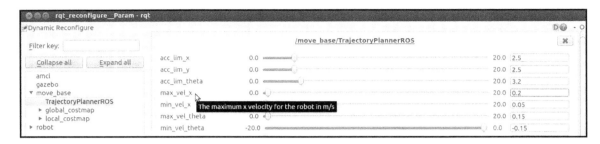

Dynamic parameter configuration

For example, we could change the `max_vel_x` or `max_vel_theta` parameters, which are configured in the `base_local_planner_params.yaml` file.

Moreover, we can view a brief description of the parameter by placing the mouse over the parameter name. This is very useful for finding out the function of each parameter.

Autonomous navigation of mobile robots – avoiding obstacles

The ROS navigation stack is the most powerful software package that allows mobile robots to move from one place to another safely. The goal of the navigation stack is to produce a safe path, in other words, a collision-free path for the robot to execute by processing the data from sensors and the environment map. One of the great functionalities of the navigation stack is obstacle avoidance. We can easily view this feature by adding an object in front of the moving robot in Gazebo.

Getting ready

A few predefined shapes or obstacles can be added in Gazebo by using the Insert model option from GUI. The navigation stack detects the obstacle and automatically creates an alternative possible collision path.

How it works...

At the same time, we can see a new global plan to avoid the obstacle in the RViz window. This feature is most interesting when the robot is moving in real environments surrounded by static and dynamic obstacles. If the robot detects a possible collision, it will change direction, and try to arrive at the goal through an alternate path.

It is important to know that the detection of such obstacles is confined into the area covered by the local planner costmap (for example, 5 x 5 meters around the robot). We can view this feature in the following screenshot:

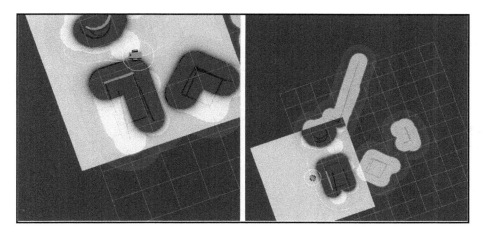

Obstacle avoidance

Sending goals

We have played around with the robot by moving it around the map using RViz and its interface. Although it is interesting and fun, it's a little tedious and not very functional.

Getting ready

Alternatively, we can program a list of waypoints and send them to the robot with only the press of a button, even when we are not in front of a computer with RViz. In this section, we will learn how to do this by using `actionlib`. We already discussed `actionlib` in Chapter 3, *ROS Architecture and Concepts - II.*

However, `actionlib` has a very similar interface to services, but if the server is taking a long time to process the request, the client can cancel the request during execution. In addition, it will also be receiving periodic feedback about how the request is progressing.

How it works...

Next, we will create a new file with the name `sendGoals.cpp` in the `chapter7_tutorials/src` folder and add the following code there:

```cpp
#include <ros/ros.h>
 #include <move_base_msgs/MoveBaseAction.h>
#include <actionlib/client/simple_action_client.h>
#include <tf/transform_broadcaster.h>
#include <sstream>
typedef actionlib::SimpleActionClient<move_base_msgs::MoveBaseAction>
MoveBaseClient;
int main(int argc, char** argv){
ros::init(argc, argv, "navigation_goals");
MoveBaseClient ac("move_base", true);
while(!ac.waitForServer(ros::Duration(5.0))){
ROS_INFO("Waiting for the move_base action server");
}
move_base_msgs::MoveBaseGoal goal;
goal.target_pose.header.frame_id = "map";
goal.target_pose.header.stamp = ros::Time::now();
goal.target_pose.pose.position.x = 1.0;
goal.target_pose.pose.position.y = 1.0;
goal.target_pose.pose.orientation.w = 1.0;
ROS_INFO("Sending goal");
ac.sendGoal(goal);
ac.waitForResult();
if(ac.getState() == actionlib::SimpleClientGoalState::SUCCEEDED)
ROS_INFO("You have arrived to the goal position");
else{
ROS_INFO("The base failed for some reason");
}
return 0;
}
```

The preceding code sample is a basic example for sending a goal to the robot.

Then, we will compile the package and launch the navigation stack to test the new program. We will need to run the following commands in two separate terminals to launch all the nodes and the configurations:

```
$ roslaunch chapter7_tutorials chapter7_configuration_gazebo.launch
$ roslaunch chapter7_tutorials move_base.launch
```

After configuring the 2D pose estimate, we will need to run the `sendGoal` node with the following command in a new terminal:

```
$ rosrun chapter6_tutorials sendGoals
```

In the RViz screen, we can observe a new global plan over the map, as shown in the following screenshot:

sendGoals

This confirms that the navigation stack has accepted the new goal command. When the robot arrives at the goal, we will see the "*You have arrived to the goal position*" message in the shell where `sendGoals` was running. In this way, we can make a list of goals or waypoints, and create a route for the robot and plan a mission.

The Robotic Arm in ROS

8

In this chapter, we will discuss the following recipes:

- Basic concepts of MoveIt!
- Motion planning using graphical interfaces
- Performing motion planning using control programs
- Adding perception to motion planning
- Grasping action with the robotic arm or manipulator

Introduction

In the previous chapter, we learned how to configure the navigation stack with any mobile robot. Subsequently, we looked into executing **Simultaneous Localization and Mapping (SLAM)**. In addition, we have also discussed the **Adaptive Monte Carlo Localization (AMCL)** algorithm of the navigation stack.

In this chapter, we will address how to create and configure a MoveIt! package for a manipulator robot and perform motion planning. We will also learn how to add perception and perform grasping.

ROS manipulation is the term used to refer to any robot that manipulates something in its environment. And what does this mean? Well, it means that it physically alters something in the world, for instance, by changing it from its initial position.

The main goal of this chapter is to learn the basic tools that we need to know about in order to understand how ROS manipulation works, and how to implement them for any manipulator robot.

There are many reasons why we should learn to perform manipulation. Manipulator robots are already being used in many environments, providing solutions to tasks that are really hard to develop for us humans, for many reasons. We will cover some of these reasons in the following subsections.

Dangerous workspaces

This refers to environments that are dangerous for humans to be in:

- Space exploration
- Foundries
- Underwater environments
- Factories

Here's a photograph of one such dangerous workspace:

Underwater operation

Repetitive or unpleasant work

Basically, this refers to industrial environments, where robots have to do repetitive tasks for prolonged periods of time:

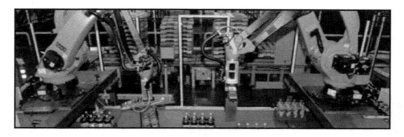

Assembly work in a factory

Human-intractable workspaces

This refers to workspaces that are very hard to be managed by humans. For instance:

- Workspaces that are too small
- Workspaces that are too big
- Workspaces where too much precision is needed

See the following photograph:

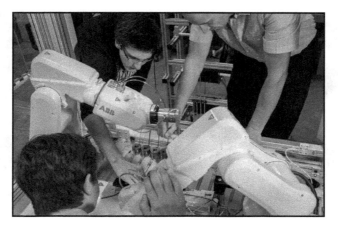

System on chip

But, there are also many other fields that are beginning to use manipulator robots. For example, surgery and patient care:

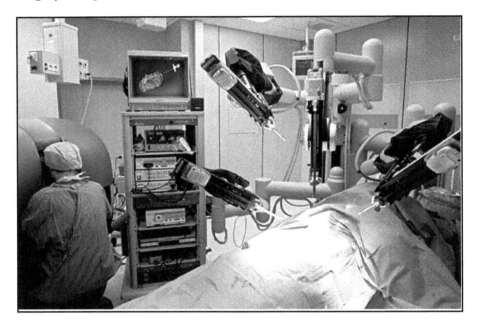

Surgical robot

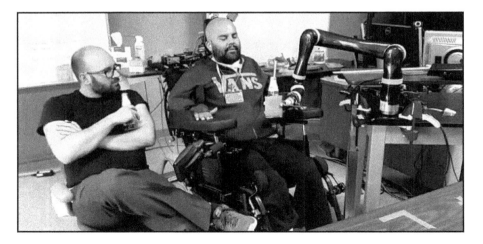

Patient care robot

Anyway, in the near future, when current limitations have been overcome, manipulation will be used in many other situations and environments. And we could be the first ones to develop such a technology!

Basic concepts of MoveIt!

This section is an introduction to some basic concepts regarding ROS manipulation that we need to know in order to be able to fully understand the contents of the chapter.

Basically, we'll need to go through the following four main topics:

- MoveIt
- Motion planning
- Perception
- Grasping

So, as we've pointed out already, the first thing we'll need to know about is MoveIt.

MoveIt

MoveIt is a set of packages and tools that allow us to perform manipulation with ROS. MoveIt provides software and tools in order to do motion planning, manipulation, perception, kinematics, collision checking, control, and navigation. Yes! It is a huge and very useful tool. We can learn more about it by checking all of its documentation on the official website (http://moveit.ros.org).

Awesome, right? We are probably a little bit confused right now, since MoveIt is a huge tool that provides lots of options. But, don't get stressed! In the next section, we will learn how to deal with MoveIt, and we'll see how this tool can help us perform manipulation in ROS.

Motion planning

Great! So, what else do we need in order to learn ROS manipulation? Well, we'll need to know how to perform motion planning! And what does motion planning mean? Well, it basically means to plan a movement (motion) from point A to point B, without colliding with anything.

In other words, you will need to be able to control the different joints and links of your robot, avoiding collisions between them or with other elements in the environment.

Perception

We'll also need to know about perception, of course! In order to interact with any object in the environment, we first need to visualize it. We need to know where it is and what it is, and that's what perception is for!

Perception is usually done using RGBD cameras such as Kinect, which is used to improve manipulation tasks such as detecting new objects that spawn into the simulation.

Grasping

Finally, you'll need to know about grasping. And what is grasping? Well, the word grasping refers to the action of catching an object from the environment in order to perform an action with it; for instance, to change an object's position. During the grasping process, there are other variables that take place, such as the perception of the environment.

Even though grasping may look like a very easy and simple task, it is not. Not at all! There are lots of variables that need to be taken into account, and there are lots of things that can go wrong! We will learn the basics about grasping in the upcoming sections.

Getting ready

We'll need a manipulator robot to perform these tasks. We know that a manipulator robot is a kind of robot that is able to physically alter the environment it works in.

A manipulator robot is modeled as a chain of rigid links, which are connected by joints, and which end in what is known as the end-effector of the robot. So, basically, any manipulator is composed of these three elemental parts:

- **Links:** Rigid pieces that connect the joints of the manipulator
- **Joints:** Connectors between the links of the manipulator, providing either translational or rotational movement
- **End Effector**: These include the following:
  - Grippers/tools
  - Grippers/tools with sensors

In the following screenshot, we can have a look at the manipulator robot and identify the names of all the links, joints, and end effectors that are part of it:

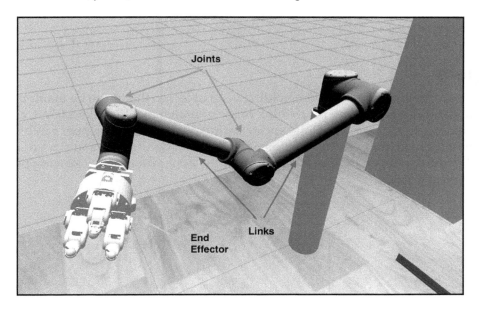

Robot manipulator (arm)

Finally, it would be good to be familiar with some basic terminology that is often used in robotic manipulation.

DoFs for manipulation

DoFs is the word used to refer to the **Degrees of Freedom** of a robot. And what does this mean? Well, it basically means the number of ways in which a robotic arm can move. A system has n DoFs if exactly n parameters are required to completely specify the configuration:

- In the case of a manipulator robot, the configuration is determined by the number of joints that the robot has. So, the number of joints determines the number of DoFs of the manipulator.
- A rigid object in a 3D space has six parameters, where there are three for positioning (x, y, z) and three for the orientation (roll, pitch, and yaw angles).

If a manipulator has fewer than six DoFs, the arm cannot reach every point in the workspace with arbitrary orientation, whereas if a manipulator has more than six DoFs, the robot is kinematically redundant. Also, the more DoFs a manipulator has, the harder it will be to control it.

Grippers

A gripper is a device that enables the holding of an object so that it can be manipulated. The easiest way to describe a gripper is to think of the human hand. Just like a hand, a gripper enables holding, tightening, handling, and releasing of an object. A gripper can be attached to a robot or it can be part of a fixed automation system. Many styles and sizes of grippers exist so that the correct model can be selected for the application. The following photograph shows a typical gripper:

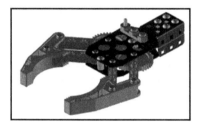

Gripper

Motion planning using graphical interfaces

In this section, we will learn how to create a MoveIt package for our industrial robot. By completing this section, we will be able to create a package that allows our robot to perform motion planning.

As we mentioned in the previous section, MoveIt is a very complex and useful tool. So, within this subsection, we are not going to dive into the details of how MoveIt works, or all the features it provides.

Fortunately, MoveIt provides a very nice and easy-to-use GUI, which will help us interact with the robot in order to perform motion planning. However, before being able to actually use MoveIt, we need to build a package. This package will generate all the configuration and launch files required for using our defined robot (the one that is defined in the URDF file) with MoveIt.

Getting ready

First of all, we'll need to launch the MoveIt setup assistant. We can do that by typing the following command:

```
$ roslaunch moveit_setup_assistant setup_assistant.launch
```

 We have to install the MoveIt ROS package if it is not installed in the default installation. Type the following command to install it:
```
$ sudo apt-get install ros-kinetic-moveit.
```

We will see something like what is shown in the following screenshot:

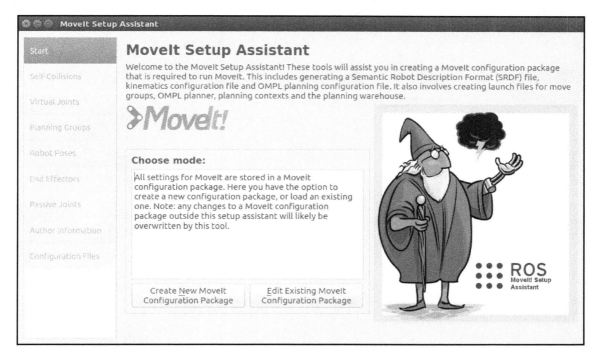

MoveIt Setup Assistant application

Great! We are now at the **MoveIt Setup Assistant**. The next thing we'll need do is load our robot file. So, let's continue!

Click on the **Create New MoveIt Configuration Package** button. A new section like this will appear:

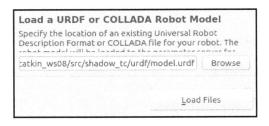

Load a URDF or COLLADA Robot Model
Specify the location of an existing Universal Robot Description Format or COLLADA file for your robot. The

catkin_ws08/src/shadow_tc/urdf/model.urdf   Browse

Load Files

Creating a new MoveIt configuration package

Now, just click the **Browse** button, select the URDF file named `model.urdf` located in the `chapter8_tutorials` package, and click on the **Load Files** button. We will probably need to copy this file into our workspace. We should now see something like this:

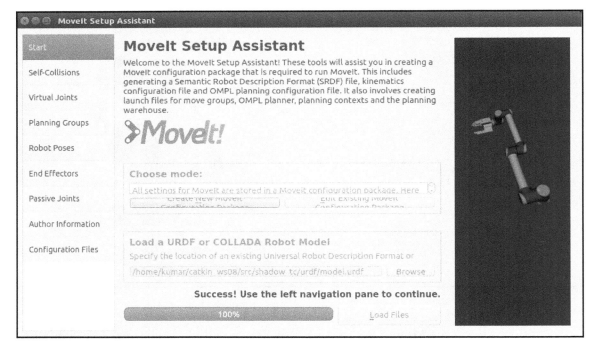

Loading a robot model

Great! We've loaded the xacro file of our robot into the **MoveIt Setup Assistant**. Now, let's start configuring this.

We will go to the **Self-Collisions** tab and click on the **Regenerate Default Collision Matrix** button. We will end up with something like this:

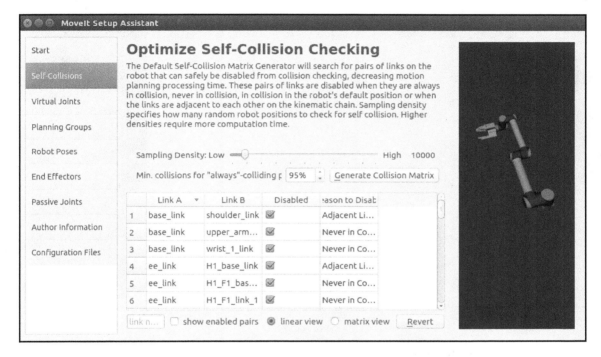

Regenerating the default collision matrix

Here, you are just defining some pairs of links that don't need to be considered when performing collision checking. For instance, because they are adjacent links, they will always be in a collision.

Next, we will move on to the **Virtual Joints** tab. Here, we will define a virtual joint for the base of the robot. Click the **Add Virtual Joint** button and set the name of this joint to **FixedBase**, and the parent to the world. Just like this:

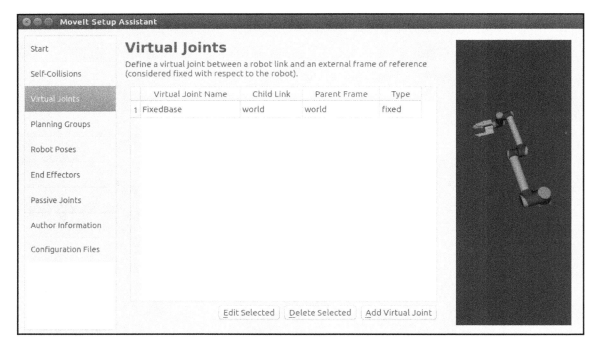

Adding a virtual joint

Finally, we will click the **Save** button. Basically, what we are doing here is creating an *imaginary* joint that will connect the base of our robot with the simulated world.

Now, we will open the **Planning Groups** tab and click the **Add Group** button. Here, we will create a new group called arm, which will use the KDLKinematicsPlugin, just like in the following screenshot:

Planning Groups

Create and edit planning groups for your robot based on joint collections, link collections, kinematic chains and subgroups.

Create New Planning Group

Group Name:	arm
Kinematic Solver:	kdl_kinematics_plugin/KDLKinematicsPlugin
Kin. Search Resolution:	0.005
Kin. Search Timeout (sec):	0.005
Kin. Solver Attempts:	3

Planning group

Next, we will click on the **Add Joints** button, and we will select all the joints that form the arm of the robot, excluding the gripper. Just like this:

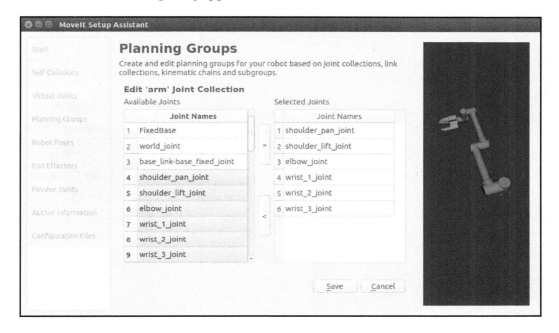

Adding joints

Finally, we will click the **Save** button, and we will end up with something like this:

Planning Groups

Create and edit planning groups for your robot based on joint collections, link collections, kinematic chains and subgroups.

Current Groups

▼ **arm**
    ▼ *Joints*
        shoulder_pan_joint - Revolute
        shoulder_lift_joint - Revolute
        elbow_joint - Revolute
        wrist_1_joint - Revolute
        wrist_2_joint - Revolute
        wrist_3_joint - Revolute
    Links
    Chain
    Subgroups

Saving the joints configuration

So now, we've defined a group of links for performing motion planning with, and we've defined the plugin we want to use to calculate those plans.

Now, we will repeat the same process, but this time for the gripper. In this case, we do not have to define any kinematics solver. If you are not sure of what joints to add to the hand, you can have a look at the following screenshot:

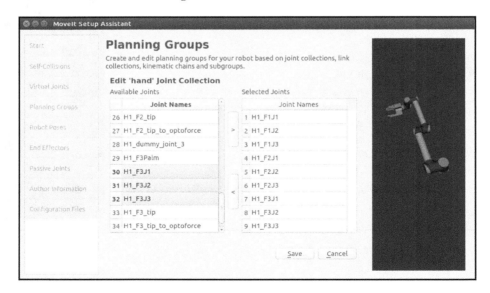

Gripper joints

At the end, we should end up with something similar to this:

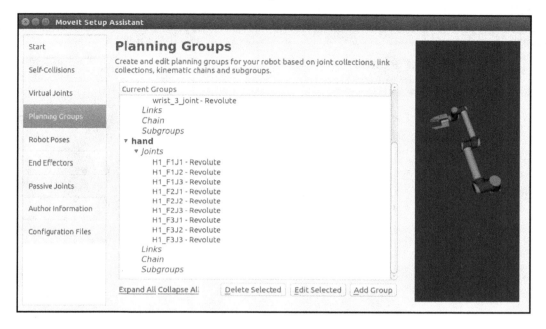

Joints configuration

Now, we are going to create a couple of predefined poses for our robot. We will go to the **Robot Poses** tab and click on the **Add Pose** button. On the left-hand side of the screen, we will be able to define the name of the pose and the planning group it refers to. In this case, we will name the first pose open, and it will be related, obviously, to the **hand** group:

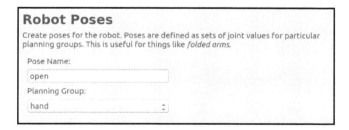

Robot poses: hand

Now, we will have to define the positions of the joints that will be related to this pose. For this case, we can set them as in the following screenshot:

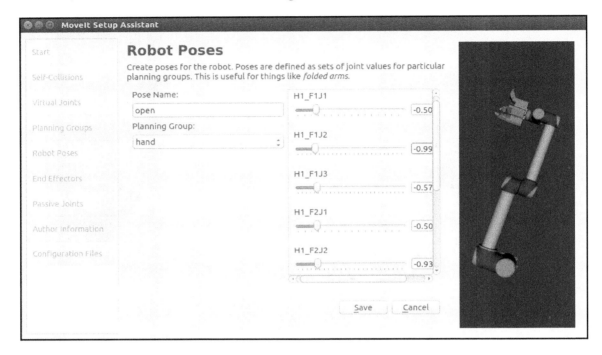

Open hand joints configuration

Now, we will repeat this operation, but this time we will define the `close` pose. For instance, it could be something like this:

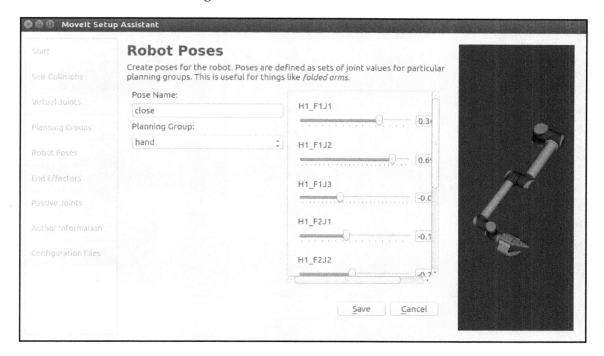

Closed hand joints configuration

Finally, let's create a `start` pose for the **arm** group. It could be something like this:

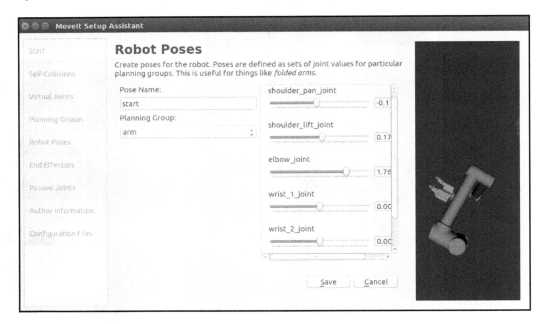

Start arm joints configuration

The next step will be to set up the End Effector of the robot. For that, just go the **End Effectors** tab, and click on the **Add End Effector** button. We will name our End Effector `hand`:

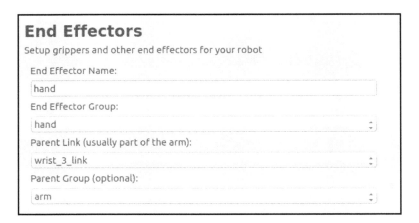

End Effector configuration

Next, we just have to enter our name and email address in the **Author Information** tab.

Finally, we will go to the **Configuration Files** tab and click the **Browse** button. Navigate to the `catkin_ws/src` directory, create a new directory, and name it `myrobot_moveit_config`. Click on **Choose** to select the directory you've just created:

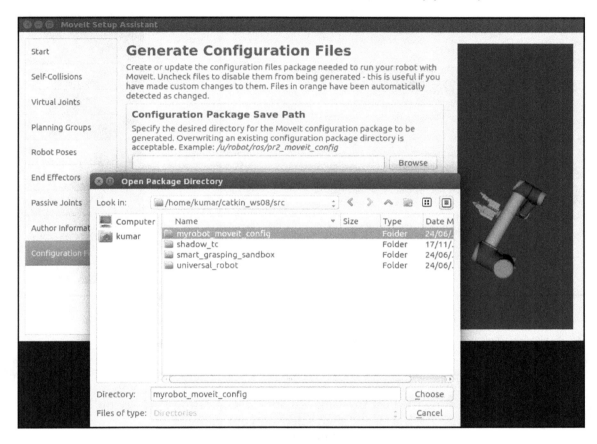

Configuration files

Now, we will click on the **Generate Package** button. If everything goes well, we should see something like this:

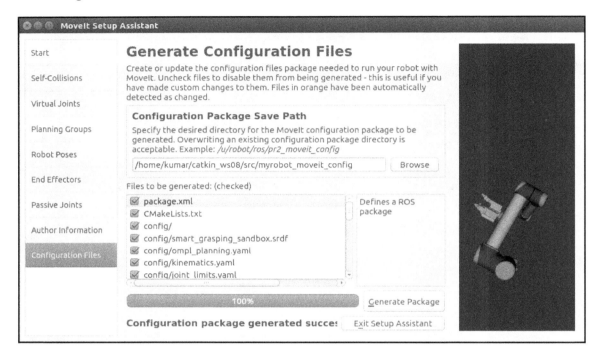

Saving configuration files

And that's it! We have just created a MoveIt package for our articulated robot.

 If, for any reason, you need to edit your MoveIt package (for instance, in future exercises when you detect that you made an error), you can do that by selecting the **Edit Existing Moveit Configuration Package** option in the Setup Assistant, and then selecting your package.
If you modify your MoveIt package, you will need to restart the simulation in order to make the appropriate changes.

Now that we've created a MoveIt package for our robot, and we've worked a little more with it, let's take a deeper look at some key aspects of Moveit.

MoveIt architecture

Let's start with a quick look at the MoveIt architecture. Understanding the architecture of MoveIt! helps to program and interface the robot to MoveIt. Here, we can have a look at the following diagram showing the MoveIt architecture:

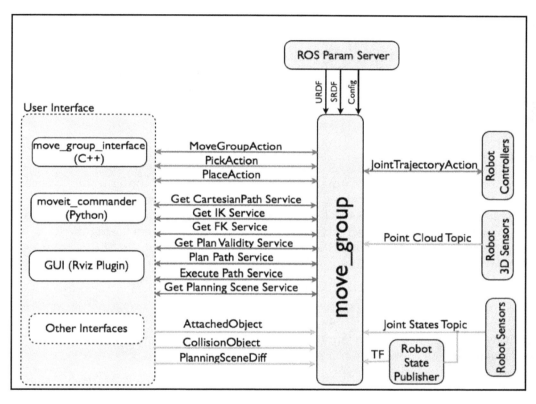

MoveIt architecture

We can say that `move_group` is the heart of MoveIt, as this node acts as an integrator of the various components of the robot and delivers actions/services according to the user's needs.

The `move_ group` node collects robot information, such as the PointCloud, the joint state of the robot, and the transforms (TFs) of the robot in the form of topics and services.

From the parameter server, it collects the robot kinematics data, such as robot description (URDF), **Semantic Robot Description Format** (**SRDF**), and the configuration files. The SRDF file and the configuration files are generated when we generate a MoveIt! package for our robot. The configuration files contain the parameter file for setting joint limits, perception, kinematics, end effector, and so on. These are the files that have been created in the config folder of your package.

How it works...

When MoveIt! gets all of this information about the robot and its configuration, we can say that it is properly configured and we can start commanding the robot from the user interfaces. We can either use C++ or Python MoveIt! APIs to command the move group node to perform actions such as pick/place, IK, or FK, among others. Using the RViz motion planning plugin, we can command the robot from the RViz GUI itself. This is what we are going to do in the next subsection!

Basic motion planning

To start, we are simply going to launch the MoveIt Rviz environment and begin doing some tests regarding motion planning.

We will execute the following command in order to start the MoveIt RViz demo environment:

```
$roslaunch myrobot_moveit_config demo.launch
```

If everything goes fine, we will see something like this:

MoveIt motion planning window

Now, we will move to the **Planning** tab, as shown in the following screenshot:

Motion planning configuration

Before we start planning anything, it is always a good practice to update the current start state. Therefore, in the **Query** section, in **Select Goal State**, we will choose the start option (which we used for one of the poses that we defined in the previous subsection) and click on the **Update** button. Our robot scene will be updated with the new position that has been selected.

Now, we can click on the **Plan** button in the **Commands** section. The robot will begin to plan a trajectory to reach that point, as shown in the following screenshot:

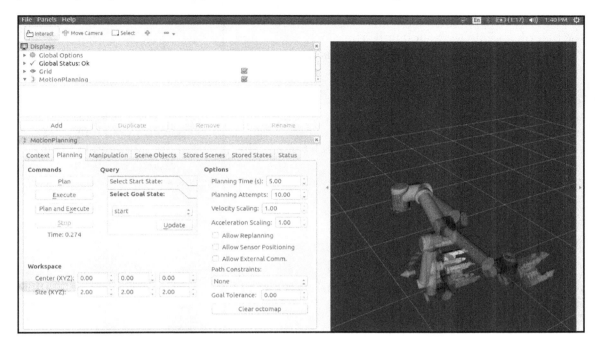

Motion planning – planning the start state

Finally, if we click on the **Execute** button, the robot will execute that trajectory.

We have just played around with the new tool! We can repeat this process again if we wish to, but with different options selected. For instance, instead of moving the robot to the start position, we could set a random valid position as the goal. We could also try to check and uncheck the different visualization options that appear in the upper **Displays** section.

How it works...

We've now seen how to perform some basic motion planning through the MoveIt RViz GUI, and we're a little more familiar with MoveIt. So, let's discuss some interesting concepts!

MoveIt! planning scene

The term *planning scene* is used to represent the world around the robot and is used to store the state of the robot itself. The planning scene monitor inside of `move_group` maintains the planning scene representation. The `move_group` node consists of another section called the world geometry monitor, which builds the world geometry from the sensors of the robot and from the user input.

The planning scene monitor reads the `joint_state` topics from the robot, and the sensor information and world geometry from the world geometry monitor. The world scene monitor reads from the occupancy map monitor, which uses 3D perception to build a 3D representation of the environment, called **Octomap**. The Octomap can be generated from PointClouds, using PointCloud occupancy map update plugin. Similarly, the depth images are generated by a depth image occupancy map updater. You will see this part in the next section when we introduce perception.

MoveIt! kinematics handling

MoveIt! provides great flexibility for switching the inverse kinematics algorithms using the robot plugins. Users can write their own IK solver as a MoveIt! plugin and switch from the default solver plugin whenever required. The default IK solver in MoveIt! is a numerical Jacobian-based solver.

Compared to the analytic solvers, the numerical solver can take more time to solve IK. The package called **IKFast** can be used to generate a C++ code for solving IK using analytical methods, which can be used for different kinds of robot manipulators and will perform better for six or fewer degrees of freedom. This C++ code can also be converted into the MoveIt! plugin by using some ROS tools.

Forward kinematics and finding jacobians are already integrated into the MoveIt! RobotState class, so we don't need to use plugins for solving FK.

MoveIt! collision checking

The `CollisionWorld` object inside MoveIt! is used to find collisions inside a planning scene, which are using the **Flexible Collision Library** (FCL) package as a backend. MoveIt! supports collision checking for different types of objects, such as meshes; primitive shapes such as boxes, cylinders, cones, spheres, and so on; and Octomap.

Collision checking is one of the most computationally expensive tasks during motion planning. To reduce this computation, MoveIt! provides a matrix called **Allowed Collision Matrix (ACM)**. It contains a binary value corresponding to the need to check for collisions between two pairs of bodies. If the value of the matrix is 1, it means that the collision of the corresponding pair is not needed. We can set the value to 1, where the bodies are always so far away that they will never collide with each other. Optimizing ACM can reduce the total computation needed for collision avoidance. This was done when we were creating the package, as you may remember!

There's more...

Until now, though, we've only moved the robot in the MoveIt application. This is very useful, because we can do many tests without worrying about any damage to robot and environment. Anyway, the final goal will always be to move the real robot.

The MoveIt package we've created is able to provide the necessary ROS services and actions in order to plan and execute trajectories, but it isn't able to pass these trajectories to the real robot. All the kinematics we've been performing were executed in an internal simulator that MoveIt provides. In order to communicate with the real robot, it will be necessary to make a couple of modifications to the MoveIt package that we created at the beginning of this section.

Obviously, we don't have a real robot to do this, so we will apply the same but for moving the simulated robot. We will look at what we need to change in our MoveIt package in the following subsection.

Moving the real robot

First of all, we'll need to create a file to define how we will control the joints of our *real* robot. Inside the `config` folder of our MoveIt package, create a new file named `controllers.yaml`. Copy the following content inside of it:

```
controller_list:
  - name: arm_controller
    action_ns: follow_joint_trajectory
    type: FollowJointTrajectory
    joints:
      - shoulder_pan_joint
      - shoulder_lift_joint
      - elbow_joint
      - wrist_1_joint
```

```
      - wrist_2_joint
      - wrist_3_joint
  - name: hand_controller
    action_ns: follow_joint_trajectory
    type: FollowJointTrajectory
    joints:
      - H1_F1J1
      - H1_F1J2
      - H1_F1J3
      - H1_F2J1
      - H1_F2J2
      - H1_F2J3
      - H1_F3J1
      - H1_F3J2
      - H1_F3J3
```

So basically, here, we are defining the action servers that we will use for controlling the joints of our robot.

First, we are setting the name of the joint trajectory controller action server for controlling the arm of the robot. And how do we know that? Well, if we do a `rostopic list` in any command-line interface, we'll find our topics with the following structure:

```
user ~ $ rostopic list
/arm_controller/command
/arm_controller/follow_joint_trajectory/cancel
/arm_controller/follow_joint_trajectory/feedback
/arm_controller/follow_joint_trajectory/goal
/arm_controller/follow_joint_trajectory/result
/arm_controller/follow_joint_trajectory/status
/arm_controller/state
```

The joint trajectory controller action server for the arm

Doing this, we know that our robot has a joint trajectory controller action server that is called `/arm_controller/follow_joint_trajectory/`. We can also find this out by checking the messages used by action server, of type `FollowJointTrajectory`.

Finally, we already know the names of the joints that our robot uses. We saw them while we were creating the MoveIt package, and we can also find them in the `model.urdf` file.

Similarly, we will simply repeat the process described just now, but for the `/hand_controller/follow_joint/trajectory` action server:

```
/hand_controller/command
/hand_controller/follow_joint_trajectory/cancel
/hand_controller/follow_joint_trajectory/feedback
/hand_controller/follow_joint_trajectory/goal
/hand_controller/follow_joint_trajectory/result
/hand_controller/follow_joint_trajectory/status
```

The joint trajectory controller action server for the hand

Next, we'll have to create a file to define the names of the joints of our robot inside the config directory; create a new file called `joint_names.yaml` and copy the following content inside of it:

```
controller_joint_names: [shoulder_pan_joint, shoulder_lift_joint,
elbow_joint, wrist_1_joint, wrist_2_joint, wrist_3_joint, H1_F1J1, H1_F1J2,
H1_F1J3, H1_F2J1, H1_F2J2, H1_F2J3, H1_F3J1, H1_F3J2, H1_F3J3]
```

Now, if we open the `smart_grasping_sandbox_moveit_controller_manager.launch.xml` file, which is inside the `launch` directory, we'll see that it's empty. Therefore, we will add the following content inside of it:

```
<launch>
  <rosparam file="$(find myrobot_moveit_config)/config/controllers.yaml"/>
  <param name="use_controller_manager" value="false"/>
  <param name="trajectory_execution/execution_duration_monitoring"
value="false"/>
  <param name="moveit_controller_manager"
value="moveit_simple_controller_manager/MoveItSimpleControllerManager"/>
</launch>
```

What we are doing here is basically loading the `controllers.yaml` file we have just created as well as the `MoveItSimpleControllerManager` plugin, which will allow us to send the plans calculated in MoveIt to our *real* robot, in this case, the simulated robot.

Finally, we will have to create a new launch file that sets up the system to control our robot. So, inside the `launch` directory, we will create a new launch file called `myrobot_planning_execution.launch` with the following contents:

```
<launch>
  <rosparam command="load" file="$(find
myrobot_moveit_config)/config/joint_names.yaml"/>

  <include file="$(find
```

```
myrobot_moveit_config)/launch/planning_context.launch" >
    <arg name="load_robot_description" value="true" />
  </include>

  <node name="joint_state_publisher" pkg="joint_state_publisher"
type="joint_state_publisher">
    <param name="/use_gui" value="false"/>
    <rosparam param="/source_list">[/joint_states]</rosparam>
  </node>

  <include file="$(find myrobot_moveit_config)/launch/move_group.launch">
    <arg name="publish_monitored_planning_scene" value="true" />
  </include>

  <include file="$(find myrobot_moveit_config)/launch/moveit_rviz.launch">
    <arg name="config" value="true"/>
  </include>
</launch>
```

Finally, we are loading the `joint_names.yaml` file and launching some launch files we need in order to set up the MoveIt environment. We can check what those launch files do if we want. For now, though, let's focus for a moment on the `joint_state_publisher` node that is being launched.

If we do again a `rostopic list`, you will see that there is a topic called `/joint_states`. It is in this topic where the states of the joints of the simulated robot are published. So, we need to put this topic into the `/source_list` parameter so that MoveIt knows where the robot is at each moment.

Finally, we just have to launch the launch file we have just created (`myrobot_planning_execution.launch`) and plan a trajectory, just as we learned to do in the previous section. Once the trajectory is planned, we can press the **Execute** button in order to execute the trajectory in the simulated robot.

Congratulations! We can see the simulated robot executing the trajectory in the following screenshot:

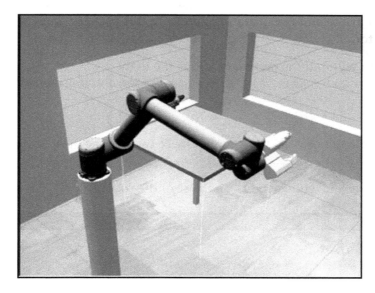

Robotic arm executing a trajectory

Performing motion planning using control programs

In the previous section, we learned that we can plan and execute trajectories for our robot using the MoveIt RViz environment. But, this is not the most common case.

Usually, we would want to move our robot with our control program, and this is exactly what we are going to do in this chapter! For this course, we are going to use Python to control the robot, because it's easier and faster.

First of all, we will have to create a MoveIt package for the fetch robot, just like we learned to do in the previous section.

Getting ready

We can create a MoveIt package for the simulated robot, just like we did in the previous section. We will have to add two planning groups to this package, one for the `arm`, and one for the end-effector.

 You will find the necessary URDF file in the `chapter8_tutorials/src/model` folder, and it is named `fetch.urdf`.

We can create as many poses as we want, but these two are the most important poses—`start` and `home`.

The following screenshot shows the `start` pose of the robot:

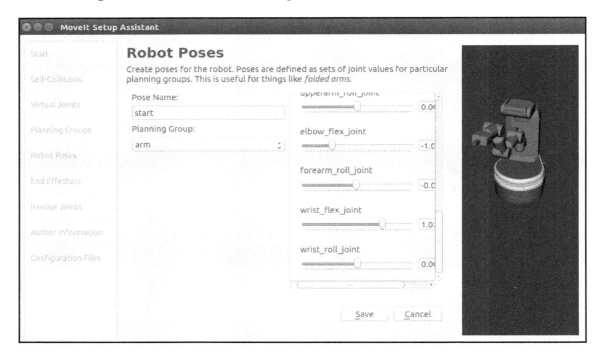

The start pose of the fetch robot

The following screenshot shows the `home` pose, which is the trivial pose on any arm robot:

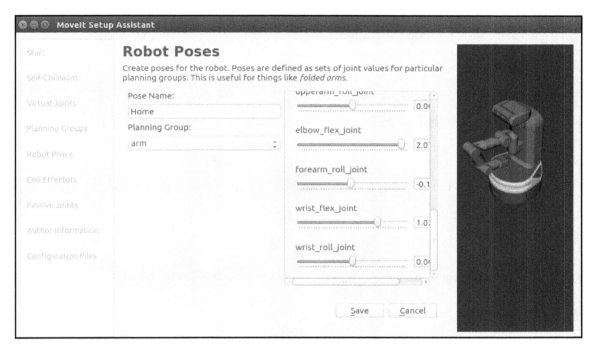

The home pose of the fetch robot

Next, we will connect this MoveIt package to the simulated robot. We will test that we can plan and execute trajectories using the MoveIt package and that these trajectories apply to the simulated robot. You may get confused when you try to add the gripper to the `controllers.yaml` file. If you know how to do it, go ahead and add it. If you don't, you can wait until next section to add it, where this will be explained. It is not mandatory to add the gripper to the `controllers.yaml` file now. It is mandatory, though, to add it into the planning groups and the end-effector sections when you create the MoveIt package. You already know how to do that! In this case, it is not required to create a Virtual Joint, so you can leave this section empty.

Great! So now that we have created the MoveIt package, we're ready to begin with the main goal of this section!

Before actually starting with the contents of this section, let's execute the following command in order to raise the fetch robot's torso. This will make it easier to plan and execute trajectories with the arm of the robot:

```
$ roslaunch fetch_gazebo_demo move_torso.launch
```

We can see how the fetch robot's torso rises in the following screenshot:

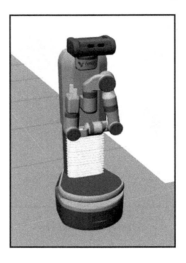

Fetch robot torso

How to do it...

As you saw in the previous section, there is a difference between planning a trajectory and executing it. In this first part of the section, we are going to see how to plan a trajectory with Python scripts.

Planning a trajectory

First of all, we'll need to launch the MoveIt RViz environment. Type the following command:

```
$ roslaunch fetch_moveit_config fetch_planning_execution.launch
```

This command may vary depending on how you've named your MoveIt package and your launch file. In this example command, it is assumed that they're named fetch_moveit_config and fetch_planning_execution.launch, respectively:

```python
#! /usr/bin/env python
import sys
import copy
import rospy
import moveit_commander
import moveit_msgs.msg
import geometry_msgs.msg
moveit_commander.roscpp_initialize(sys.argv)
rospy.init_node('move_group_python_interface_tutorial', anonymous=True)

robot = moveit_commander.RobotCommander()
scene = moveit_commander.PlanningSceneInterface()
group = moveit_commander.MoveGroupCommander("arm")
display_trajectory_publisher =
rospy.Publisher('/move_group/display_planned_path',
moveit_msgs.msg.DisplayTrajectory)

pose_target = geometry_msgs.msg.Pose()
pose_target.orientation.w = 1.0
pose_target.position.x = 0.96
pose_target.position.y = 0
pose_target.position.z = 1.18
group.set_pose_target(pose_target)
plan1 = group.plan()
rospy.sleep(5)
moveit_commander.roscpp_shutdown()
```

After a few seconds, we'll see how the robot is planning the specified motion that was described in the preceding code:

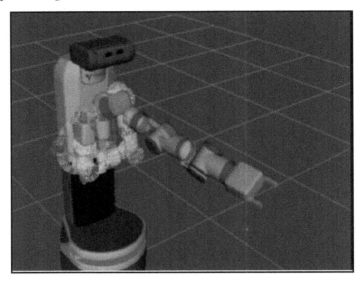

Planning a trajectory

That's great! But, how does this code work? What does each part mean? Let's break it down into smaller pieces:

```
import sys
import copy
import rospy
import moveit_commander
import moveit_msgs.msg
import geometry_msgs.msg
```

In this section of the code, we are just importing some modules and messages that we'll need for the program. The most important one here is the moveit_commander module, which will allow us to communicate with the MoveIt RViz interface:

```
moveit_commander.roscpp_initialize(sys.argv)
```

Here, we are just initializing the moveit_commander module:

```
rospy.init_node('move_group_python_interface_tutorial', anonymous=True)
```

Here, we are just initializing a ROS node:

```
robot = moveit_commander.RobotCommander()
```

Here, we are creating a `RobotCommander` object, which is, basically, an interface to our robot:

```
scene = moveit_commander.PlanningSceneInterface()
```

Here, we are creating a `PlanningSceneInterface` object, which is, basically, an interface to the world that surrounds the robot:

```
group = moveit_commander.MoveGroupCommander("arm")
```

Here, we create a `MoveGroupCommander` object, which is an interface to the manipulator group of joints that we defined when we created the MoveIt package, back in the first section of this chapter. This will allow us to interact with this set of joints, which, in this case, is the full arm:

```
display_trajectory_publisher =
rospy.Publisher('/move_group/display_planned_path',
moveit_msgs.msg.DisplayTrajectory)
```

Here, we are defining a topic `Publisher`, which will publish to the `/move_group/display_planned_path` topic. By publishing into this topic, we will be able to visualize the planned motion through the MoveIt RViz interface:

```
pose_target = geometry_msgs.msg.Pose()
pose_target.orientation.w = 1.0
pose_target.position.x = 0.7
pose_target.position.y = -0.05
pose_target.position.z = 1.1
```

Here, we are creating a pose object, which is the type of message that we will send as a goal. Then, we just give values to the variables that will define the goal pose:

```
plan1 = group.plan()
```

Finally, we are telling the manipulator group that we created previously to calculate the plan. If the plan is successfully computed, it will be displayed through MoveIt RViz:

```
moveit_commander.roscpp_shutdown()
```

Finally, we just shut down the `moveit_commander` module.

For executing the preceding code, we will create a new package called `my_motion_scripts`. Inside this package, we will create a new directory called `src`, with a file named `planning_script.py`. Finally, copy the code we've just discussed inside this file. Inside the package, we will also create a `launch` directory that contains a `launch` file in order to launch the `planning_script.py` file.

Then, we can modify the values assigned to the `pose_target` variable. Afterwards, launch our code and check if the new pose was achieved successfully. We can repeat this process and try it again with different poses.

Planning to a joint space goal

Sometimes, instead of just moving the end-effector towards a goal, we may be interested in setting a goal for a specific joint. Let's see how we can do this.

First of all, we'll need to launch the MoveIt RViz environment by using the following command:

```
$ roslaunch fetch_moveit_config fetch_planning_execution.launch
```

And, execute the following Python code:

```python
#! /usr/bin/env python
import sys
import copy
import rospy
import moveit_commander
import moveit_msgs.msg
import geometry_msgs.msg

moveit_commander.roscpp_initialize(sys.argv)
rospy.init_node('move_group_python_interface_tutorial', anonymous=True)

robot = moveit_commander.RobotCommander()
scene = moveit_commander.PlanningSceneInterface()
group = moveit_commander.MoveGroupCommander("arm")
display_trajectory_publisher =
rospy.Publisher('/move_group/display_planned_path',
moveit_msgs.msg.DisplayTrajectory)
group_variable_values = group.get_current_joint_values()
group_variable_values[5] = -1.5
group.set_joint_value_target(group_variable_values)
plan2 = group.plan()
rospy.sleep(5)
moveit_commander.roscpp_shutdown()
```

When the code finishes executing, we'll see how the robot is planning the specified motion described in the preceding code:

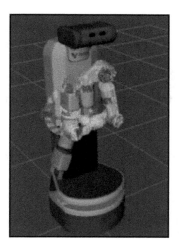

Planning to a joint space goal

That's great, right? But, as we did previously, let's analyze the new code we've introduced in order to understand what's going on:

```
group.clear_pose_targets()
```

Here, we are just clearing the actual values of the `pose_target` variable:

```
group_variable_values = group.get_current_joint_values()
```

Next, we are getting the current values of the joints:

```
group_variable_values[3] = 1.5
group.set_joint_value_target(group_variable_values)
```

Now, we will modify the value of one of the joints and set this new joint value as a target:

```
plan2 = group.plan()
```

Finally, we just compute the plan for the new joint space goal.

Inside the `my_motion_scripts` package, we will create a new file named `joint_planning.py`. Similarly, as we did previously, we can perform some tests by giving different values to different joints.

Getting useful information from motion planning

Through code, we can also get some valuable data that we may require for our code. Let's see some examples.

We can get the reference frame for a certain group by executing this line:

```
print "Reference frame: %s" % group.get_planning_frame()
```

We can get the end-effector link for a certain group by executing this line:

```
print "End effector: %s" % group.get_end_effector_link()
```

We can get a list with all of the groups of the robot, like this:

```
print "Robot Groups:"
print robot.get_group_names()
```

We can get the current values of the joints, like this:

```
print "Current Joint Values:"
print group.get_current_joint_values()
```

We can also get the current pose of the end-effector of the robot, like this:

```
print "Current Pose:"
print group.get_current_pose()
```

Finally, we can check the general status of the robot, like this:

```
print "Robot State:"
print robot.get_current_state()
```

We can create a new file inside the package called `get_data.py` and add all the new code we've learned previously and check what results we get.

Executing a trajectory

So, at this point, we've seen some methods that allow us to plan a trajectory with Python code. But, what about executing this trajectory with the real robot? In fact, it's very simple. In order to execute a trajectory, we just need to call the `go()` function from the planning group, like this:

```
group.go(wait=True)
```

By executing this line of code, we will be telling our robot to execute the last trajectory that has been set for the planning group.

First of all, we'll need to launch the MoveIt RViz environment by executing the following command:

```
$ roslaunch fetch_moveit_config fetch_planning_execution.launch
```

We will create a new Python script called `execute_trajectory.py` and copy the code from `joint_planning.py` before adding a line into our new script in order to execute that trajectory.

Moreover, we can try this with any of the codes we have created for planning trajectories. When the code finishes executing, we'll see in Gazebo how the simulated robot is planning the specified motion described in the preceding code, as shown in the following screenshot:

Executing a trajectory

 When you have finished this chapter, make sure to return the robot to the home position.

And that's it! We have finished this section! I really hope that you enjoyed it and, most of all, that you have learned a lot! Now, if you want to learn how we can add perception to motion planning tasks, just go to the next section!

Adding perception to motion planning

In the previous section, we learned how to plan and execute trajectories for our robot using programs. However, we weren't taking into account perception, were we?

Usually, we will want to take the data from a 3D vision sensor into account; for instance, from a Kinect camera. This will give us real-time information about the environment, which will allow us to plan more realistic motions, introducing any changes that the environment suffers from. So, in this section, we are going to learn how we can add a 3D vision sensor to MoveIt in order to perform vision-assisted motion planning!

Getting ready

First of all, let's make some changes to the current simulation so that we are able to work with perception better. We will create a new file named `table.urdf` in our workspace and add the following code into that file:

```
<robot name="simple_box">
  <link name="my_box">
    <inertial>
      <origin xyz="0 0 0.0145"/>
      <mass value="0.1" />
      <inertia  ixx="0.0001" ixy="0.0"  ixz="0.0"  iyy="0.0001"  iyz="0.0"
izz="0.0001" />
    </inertial>
    <visual>
      <origin xyz="-0.23 0 0.215"/>
      <geometry>
        <box size="0.47 0.46 1.3"/>
      </geometry>
    </visual>
    <collision>
      <origin xyz="-0.23 0 0.215"/>
      <geometry>
        <box size="0.47 0.46 1.3"/>
      </geometry>
    </collision>
  </link>
  <gazebo reference="my_box">
    <material>Gazebo/Wood</material>
  </gazebo>
  <gazebo>
    <static>true</static>
  </gazebo>
</robot>
```

Next, we will execute the following command in order to spawn an object right in front of the fetch robot:

```
$ rosrun gazebo_ros spawn_model -file /home/user/catkin_ws/src/table.urdf -
urdf -x 1 -model my_object
```

The output is shown in the following diagram:

Robot with an object in the environment

Later, we will execute the following command in order to move the fetch robot's head so that it points to the newly spawned object:

```
$ roslaunch fetch_gazebo_demo move_head.launch
```

We can also launch RViz and add the corresponding element in order to visualize the PointCloud of the camera, which is shown in the following screenshot:

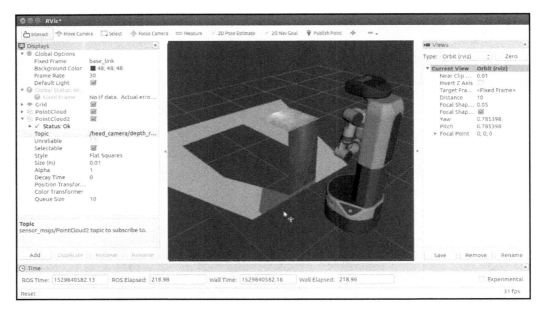

Point cloud in Rviz

How to do it...

Great! Now, we have created the appropriate environment in order to work with perception. So, let's see how we can add perception to everything we've learned about motion planning in the previous section!

Adding perception to MoveIt

In order to be able to add a sensor to the MoveIt package that we created in the previous section, we'll need to make some modifications inside the package.

First of all, we'll need to create a new file inside the config folder named `sensors_rgbd.yaml`, and inside that file, copy in the following contents:

```
sensors:
    - sensor_plugin: occupancy_map_monitor/PointCloudOctomapUpdater
      point_cloud_topic: /head_camera/depth_registered/points
```

```
max_range: 5
padding_offset: 0.01
padding_scale: 1.0
point_subsample: 1
filtered_cloud_topic: output_cloud
```

Basically, we are configuring the plugin that we'll use in order to interface the 3D sensor with MoveIt. The parameters that we are defining in the file are as follows:

- `sensor_plugin`: This parameter specifies the name of the plugin we are using in the robot.
- `point_cloud_topic`: The plugin will listen to this topic for PointCloud data.
- `max_range`: This is the distance limit, in meters, in which any points above the range will not be used for processing.
- `padding_offset`: This value will be taken into account for robot links and attached objects when filtering clouds containing the robot links (self-filtering).
- `padding_scale`: This value will also be taken into account while self-filtering.
- `point_subsample`: If the update process is slow, points can be subsampled. If we make this value greater than 1, the points will be skipped instead of processed.
- `filtered_cloud_topic`: This is the final filtered cloud topic. We will get the processed PointCloud through this topic. It is mainly used for debugging.

Next, we'll need to fill in the existing, but blank, `fetch_moveit_sensor_manager.launch.xml` file, which is located in the `launch` folder. We'll need to load the YAML file we've just created into this file:

```
<launch>
    <rosparam command="load" file="$(find
test_moveit_config)/config/sensors_rgbd.yaml" />
</launch>
```

 The content of the `rosparam` tag may vary depending on how you've named your MoveIt package. In this example command, it is assumed that the package is named `fetch_moveit_config`.

Finally, we'll have to have a look at the `sensor_manager.launch.xml` file. It should look something like this:

```
<launch>
  <!-- This file makes it easy to include the settings for sensor managers
  -->
```

```
<!-- Params for the octomap monitor -->
<!--  <param name="octomap_frame" type="string" value="some frame in
which the robot moves" /> -->
  <param name="octomap_resolution" type="double" value="0.025" />
  <param name="max_range" type="double" value="5.0" />
  <!-- Load the robot specific sensor manager; this sets the
moveit_sensor_manager ROS parameter -->
  <arg name="moveit_sensor_manager" default="fetch" />
  <include file="$(find fetch_moveit_config)/launch/$(arg
moveit_sensor_manager)_moveit_sensor_manager.launch.xml" />
</launch>
```

Similarly, the content of the `include` tag may vary depending on how we've named our MoveIt package. In this example command, it is assumed that the package is named `fetch_moveit_config`.

Now, we can launch the MoveIt RViz environment again, and we'll see a PointCloud in the scene, showing us what the robot is visualizing, as in the following screenshot:

Robot Visualization

How it works...

Interesting, right? But how does this work? What's going on internally? Let's discuss it a little bit.

Basically, we are using a plugin (`PointCloudUpdater`) that brings the simulated PointCloud obtained from the camera that is placed in fetch's head, into the MoveIt planning scene.

The robot environment is mapped as an octree representation, which can be built using a library called **OctoMap**. The OctoMap is incorporated as a plugin in MoveIt (called the Occupancy Map Updater plugin), which can update octree from different kinds of sensor inputs, such as PointClouds and depth images from 3D vision sensors.

Currently, there are the following plugins for handling 3D data in MoveIt:

- **PointCloud Occupancy Map Updater**: This plugin can take input in the form of PointClouds (`sensor_msgs/PointCloud2`). This is the one you are using in this chapter.
- **Depth Image Occupancy Map Updater**: This plugin can take input in the form of input depth images (`sensor_msgs/Image`).

There's more...

So now, we're getting real-time data from the robot's environment. Do you think that this will affect the motion plans that are calculated in any way? Let's do the following experiment in order to understand that!

First of all, we will launch the MoveIt RViz environment with perception. Next, in the motion screen, we will select the `Start` pose that we have created in the previous section, and plan a trajectory to go there, as shown in the following screenshot:

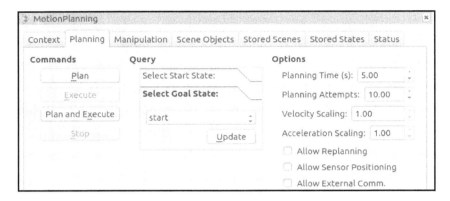

Planning a trajectory with Obstacle

Subsequently, we will remove the object from the simulation by right-clicking on it, selecting the **Delete** option, and then planning a trajectory to the start position again. Later, check if there are any differences from the previous trajectory.

That's right! The calculated motion plans will be different, depending on how the planning scene is. Makes sense, right?

We can keep on performing motion planning with the object in different positions and with different shapes, and see how it affects the trajectories that are planned. We can also execute these trajectories if we want.

See also...

Until now, you've been working using the PointCloud that is generated by the 3D sensor, which is incorporated in the fetch robot. But, we've previously seen that this is not the only way that we can add perception to MoveIt.

In the following experiment, we will see how we can also add perception without using the PointCloud.

We have to make sure that there is an object in front of the robot. If there isn't, we will execute the following command in order to spawn another object right in front of the fetch robot:

```
$ rosrun gazebo_ros spawn_model -file /home/user/catkin_ws/src/object.urdf
-urdf -x 1 -model my_object
```

Next, we have to modify the configuration files in order to use the `DepthImageUpdater` plugin, instead of the one we are currently using. As an example, we have to look at the following configuration file:

```
sensors:
    - sensor_plugin: occupancy_map_monitor/DepthImageOctomapUpdater
    image_topic: /head_camera/depth_registered/image_raw
    queue_size: 5
    near_clipping_plane_distance: 0.3
    far_clipping_plane_distance: 5.0
    skip_vertical_pixels: 1
    skip_horizontal_pixels: 1
    shadow_threshold: 0.2
    padding_scale: 4.0
    padding_offset: 0.03
    filtered_cloud_topic: output_cloud
```

Finally, we have to launch the whole environment again and plan a trajectory to check if it is detecting the environment correctly.

And that's it! We have finished this section! I really hope that you have enjoyed it and, most of all, have learned a lot! In the next section, we are going to learn grasping.

Grasping action with the robotic arm or manipulator

When working with manipulator robots, one of the main goals to achieve is picking up an object from one position, and placing it in another one, which is commonly known as **pick and place**. We call grasping the process of picking an object up by the robot an end-effector (which can be a hand, a gripper, and so on). Although this may sound like a very simple task, it is not. In fact, it's a very complex process, because lots of variables need to be taken into account when picking up the object.

We, humans, perform grasping by hand using our intelligence, but in a robot, we have to create rules for it. One of the constraints in grasping is force; the gripper/end-effector should adjust the grasping force for picking up the object, but not deform the object while grasping it.

Getting ready

First of all, we are going to see how to perform grasping in MoveIt. This means we are going to perform it without controlling the real robot.

For this, we are going to be using the `moveit_simple_grasps` package. This package is very useful for generating grasp poses for simple objects, such as blocks or cylinders, and it's quite simple, which helps in the process of learning. This package takes the position of the object to be grasped as input, and generates the necessary grasping sequences in order to pick the object up.

This package already supports robots such as Baxter, REEM, and Clam arm, among others, but it is quite easy to interface any other custom manipulator robot to the package, without having to modify too many of the codes.

First of all, we will create a new package called `myrobot_grasping`. Inside this package, create a new folder named `launch`, which will contain a launch file called `grasp_generator_server.launch`. Copy the following content inside of this file:

```
<launch>
  <arg name="robot" default="fetch"/>
  <arg name="group"        default="arm"/>
  <arg name="end_effector" default="gripper"/>
  <node pkg="moveit_simple_grasps" type="moveit_simple_grasps_server"
name="moveit_simple_grasps_server">
    <param name="group"        value="$(arg group)"/>
    <param name="end_effector" value="$(arg end_effector)"/>
    <rosparam command="load" file="$(find fetch_gazebo)/config/$(arg
robot)_grasp_data.yaml"/>
  </node>
</launch>
```

Basically, in this `launch` file, we are starting a grasp server that will provide grasp sequences to a grasp client node. We need to specify the following in the node:

- The planning group of the arm
- The planning group of the end-effector

We are also going to load the `fetch_grasp_data.yaml` file, which will contain detailed information about the gripper. We are going to create this file now.

Inside this package, we will create a new folder named `config`, and create a new file named `fetch_grasp_data.yaml` inside of it before copying the following contents into the file:

```
base_link: 'base_link'
gripper:
  #The end effector name for grasping
  end_effector_name: 'gripper'
  # Gripper joints
  joints: ['l_gripper_finger_joint', 'r_gripper_finger_joint']
  #Posture of grippers before grasping
  pregrasp_posture: [0.048, 0.048]
  pregrasp_time_from_start: 4.0
  grasp_posture: [0.016, 0.016]
  grasp_time_from_start: 4.0
  postplace_time_from_start: 4.0
  # Desired pose from end effector to grasp [x, y, z] + [R, P, Y]
  grasp_pose_to_eef: [-0.12, 0.0, 0.0]
  grasp_pose_to_eef_rotation: [0.0, 0.0, 0.0]
  end_effector_parent_link: 'wrist_roll_link'
```

So, in this file, we are basically providing detailed information about the gripper. These parameters can be tuned in order to improve the grasping process. The most important parameters that we are defining here are as follows:

- `end_effector_name`: The name of the end-effector group, as stated in the MoveIt package
- `joints`: The joints that form the gripper
- `pregrasp_posture`: The position of the gripper before grasping (open position)
- `pregrasp_time_from_start`: The time to wait before grasping
- `grasp_posture`: The position of the gripper when grasping (closed position)
- `grasp_time_from_start`: The time to wait after grasping
- `postplace_time_from_start`: The name of the end-effector
- `grasp_pose_to_eef`: The desired pose from end-effector to grasp—[x, y, z]
- `grasp_pose_to_eef`: The desired pose from end-effector to grasp—[roll, pitch, yaw]
- `end_effector_parent_link`: The name of the link that connects the gripper to the robot

So, we have now created the basic structure to create a grasping server, which will allow us to send object positions to this server, and it will provide us with a sequence so that we can grasp the specified object.

But, for that, we need to communicate with that server, and that's exactly what we are going to learn in the next subsection.

Creating a pick and place task

There are various ways in which we can perform pick and place tasks. For example, we could send the robot a predefined sequence of joint values directly, which will cause the robot to always perform the same predefined motion. In this case, we must always place the object in the same position. This method is called forward kinematics, because we have to know in advance the sequence of joint values,   in order to perform a certain trajectory.

Another method would be by using **inverse kinematics** (**IK**), without any vision feedback. In this case, we provide the robot with the pose (x, y, or z) where the object to pick is, and by doing some IK calculations, the robot will know which motions it needs to perform in order to reach the object's pose.

Similarly, another method would be to use inverse kinematics with vision support or feedback. In this case, we would use a node that would identify the pose of the object by reading the data from the sensors of the robot, such as a Kinect camera. This vision node would provide us with the pose of the object, and again, by using IK calculations, the robot would know the required motions to perform in order to reach the object.

We are probably thinking that the most efficient and stylish method is the third one, and we're right. But, this method will need object recognition knowledge, which is not the subject of this course. So, for this course, we are going to use the second method, which in the end is the same as the third one, but without the object recognition part. If you already have some perception knowledge, you can freely add an object recognition node that sends the pose of the object without having to set this pose in the code itself.

So, to summarize, what we are going to do in the next subsection of this chapter is create a node that will send an object pose to the Grasping Server, in order to receive the appropriate joint values to reach that object.

How to do it...

Inside the package that we have created in the previous subsection, we will create a new folder named scripts and create a new file named `pick_and_place.py` there before copying the following code inside of it:

```python
#!/usr/bin/env python

import rospy

from moveit_commander import RobotCommander, PlanningSceneInterface
from moveit_commander import roscpp_initialize, roscpp_shutdown

from actionlib import SimpleActionClient, GoalStatus

from geometry_msgs.msg import Pose, PoseStamped, PoseArray, Quaternion
from moveit_msgs.msg import PickupAction, PickupGoal
from moveit_msgs.msg import PlaceAction, PlaceGoal
from moveit_msgs.msg import PlaceLocation
from moveit_msgs.msg import MoveItErrorCodes
from moveit_simple_grasps.msg import GenerateGraspsAction,
GenerateGraspsGoal, GraspGeneratorOptions

from tf.transformations import quaternion_from_euler

import sys
import copy
import numpy


# Create dict with human readable MoveIt! error codes:
moveit_error_dict = {}
for name in MoveItErrorCodes.__dict__.keys():
    if not name[:1] == '_':
        code = MoveItErrorCodes.__dict__[name]
        moveit_error_dict[code] = name


class Pick_Place:
    def __init__(self):
        # Retrieve params:
        self._table_object_name = rospy.get_param('~table_object_name',
'Grasp_Table')
        self._grasp_object_name = rospy.get_param('~grasp_object_name',
'Grasp_Object')

        self._grasp_object_width = rospy.get_param('~grasp_object_width',
```

```
0.01)

        self._arm_group     = rospy.get_param('~arm', 'arm')
        self._gripper_group = rospy.get_param('~gripper', 'gripper')

        self._approach_retreat_desired_dist =
rospy.get_param('~approach_retreat_desired_dist', 0.2)
        self._approach_retreat_min_dist =
rospy.get_param('~approach_retreat_min_dist', 0.1)

        # Create (debugging) publishers:
        self._grasps_pub = rospy.Publisher('grasps', PoseArray,
queue_size=1, latch=True)
        self._places_pub = rospy.Publisher('places', PoseArray,
queue_size=1, latch=True)

        # Create planning scene and robot commander:
        self._scene = PlanningSceneInterface()
        self._robot = RobotCommander()

        rospy.sleep(1.0)

        # Clean the scene:
        self._scene.remove_world_object(self._table_object_name)
        self._scene.remove_world_object(self._grasp_object_name)

        # Add table and Coke can objects to the planning scene:
        self._pose_table    = self._add_table(self._table_object_name)
        self._pose_coke_can =
self._add_grasp_block_(self._grasp_object_name)

        rospy.sleep(1.0)

        # Define target place pose:
        self._pose_place = Pose()

        self._pose_place.position.x = self._pose_coke_can.position.x
        self._pose_place.position.y = self._pose_coke_can.position.y - 0.10
        self._pose_place.position.z = self._pose_coke_can.position.z + 0.08

        self._pose_place.orientation =
Quaternion(*quaternion_from_euler(0.0, 0.0, 0.0))

        # Retrieve groups (arm and gripper):
        self._arm    = self._robot.get_group(self._arm_group)
        self._gripper = self._robot.get_group(self._gripper_group)

        # Create grasp generator 'generate' action client:
```

```
        self._grasps_ac =
SimpleActionClient('/moveit_simple_grasps_server/generate',
GenerateGraspsAction)
        if not self._grasps_ac.wait_for_server(rospy.Duration(5.0)):
            rospy.logerr('Grasp generator action client not available!')
            rospy.signal_shutdown('Grasp generator action client not
available!')
            return

        # Create move group 'pickup' action client:
        self._pickup_ac = SimpleActionClient('/pickup', PickupAction)
        if not self._pickup_ac.wait_for_server(rospy.Duration(5.0)):
            rospy.logerr('Pick up action client not available!')
            rospy.signal_shutdown('Pick up action client not available!')
            return

        # Create move group 'place' action client:
        self._place_ac = SimpleActionClient('/place', PlaceAction)
        if not self._place_ac.wait_for_server(rospy.Duration(5.0)):
            rospy.logerr('Place action client not available!')
            rospy.signal_shutdown('Place action client not available!')
            return

        # Pick Coke can object:
        while not self._pickup(self._arm_group, self._grasp_object_name,
self._grasp_object_width):
            rospy.logwarn('Pick up failed! Retrying ...')
            rospy.sleep(1.0)

    rospy.loginfo('Pick up successfully')

        # Place Coke can object on another place on the support surface
(table):
        while not self._place(self._arm_group, self._grasp_object_name,
self._pose_place):
            rospy.logwarn('Place failed! Retrying ...')
            rospy.sleep(1.0)

    rospy.loginfo('Place successfully')

    def __del__(self):
        # Clean the scene:
        self._scene.remove_world_object(self._grasp_object_name)
        self._scene.remove_world_object(self._table_object_name)

    def _add_table(self, name):
        p = PoseStamped()
        p.header.frame_id = self._robot.get_planning_frame()
```

```
        p.header.stamp = rospy.Time.now()

        p.pose.position.x = 1.0
        p.pose.position.y = 0.08
        p.pose.position.z = 1.02

        q = quaternion_from_euler(0.0, 0.0, numpy.deg2rad(90.0))
        p.pose.orientation = Quaternion(*q)

        # Table size from ~/.gazebo/models/table/model.sdf, using the
values
        # for the surface link.
        self._scene.add_box(name, p, (0.86, 0.86, 0.02))

        return p.pose

    def _add_grasp_block_(self, name):
        p = PoseStamped()
        p.header.frame_id = self._robot.get_planning_frame()
        p.header.stamp = rospy.Time.now()

        p.pose.position.x = 0.62
        p.pose.position.y = 0.21
        p.pose.position.z = 1.07

        q = quaternion_from_euler(0.0, 0.0, 0.0)
        p.pose.orientation = Quaternion(*q)

        # Coke can size from ~/.gazebo/models/coke_can/meshes/coke_can.dae,
        # using the measure tape tool from meshlab.
        # The box is the bounding box of the coke cylinder.
        # The values are taken from the cylinder base diameter and height.
        self._scene.add_box(name, p, (0.077, 0.077, 0.070))

        return p.pose

    def _generate_grasps(self, pose, width):
        """
        Generate grasps by using the grasp generator generate action; based
on
        server_test.py example on moveit_simple_grasps pkg.
        """

        # Create goal:
        goal = GenerateGraspsGoal()

        goal.pose  = pose
        goal.width = width
```

```
        options = GraspGeneratorOptions()
        # simple_graps.cpp doesn't implement GRASP_AXIS_Z!
        #options.grasp_axis      = GraspGeneratorOptions.GRASP_AXIS_Z
        options.grasp_direction = GraspGeneratorOptions.GRASP_DIRECTION_UP
        options.grasp_rotation  = GraspGeneratorOptions.GRASP_ROTATION_FULL

        # @todo disabled because it works better with the default options
        #goal.options.append(options)

        # Send goal and wait for result:
        state = self._grasps_ac.send_goal_and_wait(goal)
        if state != GoalStatus.SUCCEEDED:
            rospy.logerr('Grasp goal failed!: %s' %
self._grasps_ac.get_goal_status_text())
            return None

        grasps = self._grasps_ac.get_result().grasps

        # Publish grasps (for debugging/visualization purposes):
        self._publish_grasps(grasps)

        return grasps

    def _generate_places(self, target):
        """
        Generate places (place locations), based on
        https://github.com/davetcoleman/baxter_cpp/blob/hydro-devel/
        baxter_pick_place/src/block_pick_place.cpp
        """

        # Generate places:
        places = []
        now = rospy.Time.now()
        for angle in numpy.arange(0.0, numpy.deg2rad(360.0),
numpy.deg2rad(1.0)):
            # Create place location:
            place = PlaceLocation()

            place.place_pose.header.stamp = now
            place.place_pose.header.frame_id =
self._robot.get_planning_frame()

            # Set target position:
            place.place_pose.pose = copy.deepcopy(target)

            # Generate orientation (wrt Z axis):
            q = quaternion_from_euler(0.0, 0.0, angle )
            place.place_pose.pose.orientation = Quaternion(*q)
```

```
        # Generate pre place approach:
        place.pre_place_approach.desired_distance =
self._approach_retreat_desired_dist
        place.pre_place_approach.min_distance =
self._approach_retreat_min_dist

        place.pre_place_approach.direction.header.stamp = now
        place.pre_place_approach.direction.header.frame_id =
self._robot.get_planning_frame()

        place.pre_place_approach.direction.vector.x =  0
        place.pre_place_approach.direction.vector.y =  0
        place.pre_place_approach.direction.vector.z = 0.2

        # Generate post place approach:
        place.post_place_retreat.direction.header.stamp = now
        place.post_place_retreat.direction.header.frame_id =
self._robot.get_planning_frame()

        place.post_place_retreat.desired_distance =
self._approach_retreat_desired_dist
        place.post_place_retreat.min_distance =
self._approach_retreat_min_dist

        place.post_place_retreat.direction.vector.x = 0
        place.post_place_retreat.direction.vector.y = 0
        place.post_place_retreat.direction.vector.z = 0.2

        # Add place:
        places.append(place)

    # Publish places (for debugging/visualization purposes):
    self._publish_places(places)

    return places

def _create_pickup_goal(self, group, target, grasps):
    """
    Create a MoveIt! PickupGoal
    """

    # Create goal:
    goal = PickupGoal()

    goal.group_name  = group
    goal.target_name = target

    goal.possible_grasps.extend(grasps)
```

```
            goal.allowed_touch_objects.append(target)

            goal.support_surface_name = self._table_object_name

            # Configure goal planning options:
            goal.allowed_planning_time = 7.0

            goal.planning_options.planning_scene_diff.is_diff = True
            goal.planning_options.planning_scene_diff.robot_state.is_diff =
    True
            goal.planning_options.plan_only = False
            goal.planning_options.replan = True
            goal.planning_options.replan_attempts = 20

            return goal

        def _create_place_goal(self, group, target, places):
            """
            Create a MoveIt! PlaceGoal
            """

            # Create goal:
            goal = PlaceGoal()

            goal.group_name           = group
            goal.attached_object_name = target

            goal.place_locations.extend(places)

            # Configure goal planning options:
            goal.allowed_planning_time = 7.0

            goal.planning_options.planning_scene_diff.is_diff = True
            goal.planning_options.planning_scene_diff.robot_state.is_diff =
    True
            goal.planning_options.plan_only = False
            goal.planning_options.replan = True
            goal.planning_options.replan_attempts = 20

            return goal

        def _pickup(self, group, target, width):
            """
            Pick up a target using the planning group
            """

            # Obtain possible grasps from the grasp generator server:
            grasps = self._generate_grasps(self._pose_coke_can, width)
```

```
        # Create and send Pickup goal:
        goal = self._create_pickup_goal(group, target, grasps)

        state = self._pickup_ac.send_goal_and_wait(goal)
        if state != GoalStatus.SUCCEEDED:
            rospy.logerr('Pick up goal failed!: %s' %
self._pickup_ac.get_goal_status_text())
            return None

        result = self._pickup_ac.get_result()

        # Check for error:
        err = result.error_code.val
        if err != MoveItErrorCodes.SUCCESS:
            rospy.logwarn('Group %s cannot pick up target %s!: %s' %
(group, target, str(moveit_error_dict[err])))

            return False

        return True

    def _place(self, group, target, place):
        """
        Place a target using the planning group
        """

        # Obtain possible places:
        places = self._generate_places(place)

        # Create and send Place goal:
        goal = self._create_place_goal(group, target, places)

        state = self._place_ac.send_goal_and_wait(goal)
        if state != GoalStatus.SUCCEEDED:
            rospy.logerr('Place goal failed!: %s' %
self._place_ac.get_goal_status_text())
            return None

        result = self._place_ac.get_result()

        # Check for error:
        err = result.error_code.val
        if err != MoveItErrorCodes.SUCCESS:
            rospy.logwarn('Group %s cannot place target %s!: %s' % (group,
target, str(moveit_error_dict[err])))

            return False
```

```python
        return True

    def _publish_grasps(self, grasps):
        """
        Publish grasps as poses, using a PoseArray message
        """

        if self._grasps_pub.get_num_connections() > 0:
            msg = PoseArray()
            msg.header.frame_id = self._robot.get_planning_frame()
            msg.header.stamp = rospy.Time.now()

            for grasp in grasps:
                p = grasp.grasp_pose.pose

                msg.poses.append(Pose(p.position, p.orientation))

            self._grasps_pub.publish(msg)

    def _publish_places(self, places):
        """
        Publish places as poses, using a PoseArray message
        """

        if self._places_pub.get_num_connections() > 0:
            msg = PoseArray()
            msg.header.frame_id = self._robot.get_planning_frame()
            msg.header.stamp = rospy.Time.now()

            for place in places:
                msg.poses.append(place.place_pose.pose)

            self._places_pub.publish(msg)


def main():
    p = Pick_Place()

    rospy.spin()

if __name__ == '__main__':
    roscpp_initialize(sys.argv)
    rospy.init_node('pick_and_place')

    main()

    roscpp_shutdown()
```

Yes, we know. There is a lot of code! And there's no explanation at all! But, don't worry if you don't understand much of the code at this point. After completing this experiment, we are going to go into detail about the main parts of the code.

First of all, we will launch the `demo.launch` file that was generated in the MoveIt package in the previous section:

```
$ roslaunch myrobot_moveit_config demo.launch
```

Afterwards, we will set the **Planning Attempts** option to something like 7, as shown in the following screenshot:

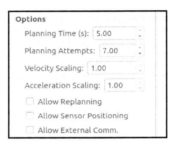

Planning Attempts

Next, we will launch the grasping server that we created in the previous subsection:

```
$ roslaunch myrobot_grasping grasp_generator_server.launch
```

Finally, we will launch the Python script that we just created, and move to the MoveIt RViz screen in order to visualize what's going on, which is shown in the screenshots in this section, in sequence:

```
$ rosrun myrobot_grasping pick_and_place.py
```

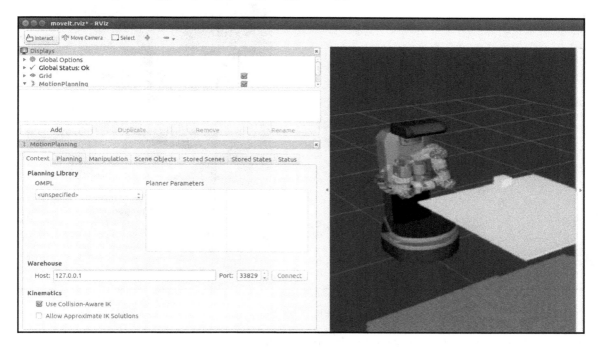

Initial status

In the preceding screenshot, the robot is shown in the initial state, which would be the home or start state:

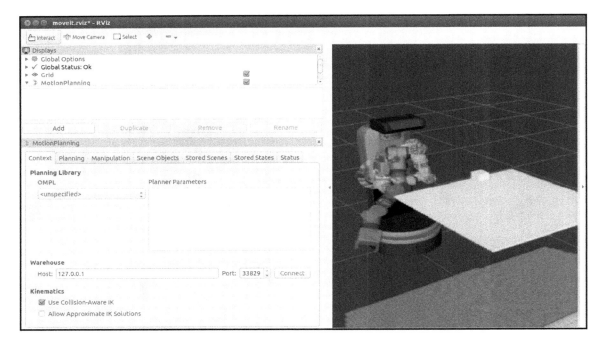

Going to grasp a pose

The preceding screenshot shows how the robot is going to grasp a pose, where it will begin grasping. Consequently, the following screenshot shows that the robot has grasped the object, which is a cube in this case:

Grasping the object

How it works...

Several things happened in this experiment, so let's discuss it in parts!

First of all, we saw a couple of cubic and rectangular blocks appearing in the RViz. Well, those were the table and the object to be grasped, of course! Inside the `pick_and_place.py` script, we created these two objects and added them into the MoveIt planning scene. Whenever we add an object to the planning scene, it will appear in this way.

Second, the pick action starts. After getting the grasp object position, our node sends this position to the grasp server, which will generate IK and check if there's any valid IK in order to pick the object up. If it finds any feasible IK solution, the arm will begin to execute the specified motions in order to pick the object up.

After the object is picked by the gripper, the place action starts. Again, our node will send a pose to the grasp server where the arm should place the object. Then, the server will check for a valid IK solution for that specified pose. If it finds any valid solution, the gripper will move to that position and release the object.

We can also have a look at the /grasp and /place topics in order to better see what's going on.

Ok, so now we have a better understanding of what happened in the previous exercise. Let's take a deeper look into the code in order to see how all of this works.

First of all, let's check how we created and added both the table and the grasping object:

```
def _add_table(self, name):
    p = PoseStamped()
    p.header.frame_id = self._robot.get_planning_frame()
    p.header.stamp = rospy.Time.now()
    #Table position
    p.pose.position.x = 0.45
    p.pose.position.y = 0.0
    p.pose.position.z = 0.22
    q = quaternion_from_euler(0.0, 0.0, numpy.deg2rad(90.0))
    p.pose.orientation = Quaternion(*q)
    # Table size
    self._scene.add_box(name, p, (0.5, 0.4, 0.02))
    return p.pose
```

In this section of the code, we are creating and adding a table to the planning scene.

Here, we first create a PoseStamped message, and we fill it in with the necessary information—the frame_id, the time stamp, and the position and orientation where the table will be placed. Then, we add the table to the scene using the add_box function.

The following code is the creation of the grasp object:

```
def _add_grasp_block_(self, name):
    p = PoseStamped()
    p.header.frame_id = self._robot.get_planning_frame()
    p.header.stamp = rospy.Time.now()
    p.pose.position.x = 0.25
    p.pose.position.y = 0.05
```

```
p.pose.position.z = 0.32
q = quaternion_from_euler(0.0, 0.0, 0.0)
p.pose.orientation = Quaternion(*q)
# Grasp Object can size
self._scene.add_box(name, p, (0.03, 0.03, 0.09))
return p.pose
```

In the preceding section of the code, we are doing the exact same thing as we did with the table, but for the `grasp` object.

After creating the `grasp` object and the grasp table, we will see how to set the pick position and the place position from the following code snippet. Here, the pose of the `grasp` object created in the planning scene is retrieved and fed into the place pose in which the Y axis of the place pose is decreased by 0.06. So, when the pick and place happens, the `grasp` object will be placed 0.06 meters (6 cm) away from the initial pose of the object in the direction of Y:

```
# Add table and grap object to the planning scene:
self._pose_table    = self._add_table(self._table_object_name)
self._pose_grasp_obj = self._add_grasp_block_(self._grasp_object_name)
rospy.sleep(1.0)
# Define target place pose:
self._pose_place = Pose()
self._pose_place.position.x = self._pose_grasp_obj.position.x
self._pose_place.position.y = self._pose_grasp_obj.position.y - 0.06
self._pose_place.position.z = self._pose_grasp_obj.position.z
self._pose_place.orientation = Quaternion(*quaternion_from_euler(0.0, 0.0, 0.0))
```

The next step is to generate the grasp pose array data for visualization, and then send the grasp goal to the grasp server. If there is a grasp sequence, it will be published; otherwise, it will respond with an error:

```
def _generate_grasps(self, pose, width):
    # Create goal:
    goal = GenerateGraspsGoal()
    goal.pose  = pose
    goal.width = width
    ....................
    ....................
    state = self._grasps_ac.send_goal_and_wait(goal)
    if state != GoalStatus.SUCCEEDED:
        rospy.logerr('Grasp goal failed!: %s' %
self._grasps_ac.get_goal_status_text())
        return None
    grasps = self._grasps_ac.get_result().grasps
```

```
    # Publish grasps (for debugging/visualization purposes):
    self._publish_grasps(grasps)
    return grasps
```

This function will create a pose array data for the pose of the place:

```
def _generate_places(self, target):
    # Generate places:
    places = []
    now = rospy.Time.now()
    for angle in numpy.arange(0.0, numpy.deg2rad(360.0),
numpy.deg2rad(1.0)):
        # Create place location:
        place = PlaceLocation()
        .........................................
        .......................................
        # Add place:
        places.append(place)
    # Publish places (for debugging/visualization purposes):
    self._publish_places(places)
```

The next function is `_create_pickup_goal()`, which will create a goal for picking up the grasping object. This goal has to be sent into MoveIt!:

```
def _create_pickup_goal(self, group, target, grasps):
    """
    Create a MoveIt! PickupGoal
    """

    # Create goal:
    goal = PickupGoal()

    goal.group_name  = group
    goal.target_name = target

    goal.possible_grasps.extend(grasps)

    goal.allowed_touch_objects.append(target)

    goal.support_surface_name = self._table_object_name

    # Configure goal planning options:
    goal.allowed_planning_time = 7.0

    goal.planning_options.planning_scene_diff.is_diff = True
    goal.planning_options.planning_scene_diff.robot_state.is_diff = True
    goal.planning_options.plan_only = False
    goal.planning_options.replan = True
```

```
goal.planning_options.replan_attempts = 20

return goal
```

Also, there is the `_create_place_goal()` function, which creates a place goal for MoveIt:

```python
def _create_place_goal(self, group, target, places):
    """
    Create a MoveIt! PlaceGoal
    """

    # Create goal:
    goal = PlaceGoal()

    goal.group_name          = group
    goal.attached_object_name = target

    goal.place_locations.extend(places)

    # Configure goal planning options:
    goal.allowed_planning_time = 7.0

    goal.planning_options.planning_scene_diff.is_diff = True
    goal.planning_options.planning_scene_diff.robot_state.is_diff = True
    goal.planning_options.plan_only = False
    goal.planning_options.replan = True
    goal.planning_options.replan_attempts = 20
    return goal
```

The important functions that are performing picking and placing are as follows. These functions will generate a pick and place sequence, which will be sent to MoveIt, and print the results of the motion planning, regardless of whether it is successful or not:

```python
def _pickup(self, group, target, width):
    """
    Pick up a target using the planning group
    """

    # Obtain possible grasps from the grasp generator server:
    grasps = self._generate_grasps(self._pose_coke_can, width)

    # Create and send Pickup goal:
    goal = self._create_pickup_goal(group, target, grasps)

    state = self._pickup_ac.send_goal_and_wait(goal)
    if state != GoalStatus.SUCCEEDED:
        rospy.logerr('Pick up goal failed!: %s' %
```

```
self._pickup_ac.get_goal_status_text())
        return None

    result = self._pickup_ac.get_result()

    # Check for error:
    err = result.error_code.val
    if err != MoveItErrorCodes.SUCCESS:
        rospy.logwarn('Group %s cannot pick up target %s!: %s' % (group,
target, str(moveit_error_dict[err])))

        return False

    return True

def _place(self, group, target, place):
    """
    Place a target using the planning group
    """

    # Obtain possible places:
    places = self._generate_places(place)

    # Create and send Place goal:
    goal = self._create_place_goal(group, target, places)

    state = self._place_ac.send_goal_and_wait(goal)
    if state != GoalStatus.SUCCEEDED:
        rospy.logerr('Place goal failed!: %s' %
self._place_ac.get_goal_status_text())
        return None

    result = self._place_ac.get_result()

    # Check for error:
    err = result.error_code.val
    if err != MoveItErrorCodes.SUCCESS:
        rospy.logwarn('Group %s cannot place target %s!: %s' % (group,
target, str(moveit_error_dict[err])))

        return False

    return True
```

See also

So, now we've seen how to perform grasping in MoveIt. But, what about applying this to the real robot? Well, actually, it's quite easy at this point.

First of all, we will execute the following commands in order to spawn a table and a grasping object into the simulation:

```
$ rosrun gazebo_ros spawn_model –database table –gazebo –model table –x
1.30 –y 0 –z 0
$ rosrun gazebo_ros spawn_model –database demo_cube –gazebo –model
grasp_cube –x 0.65 –y 0.15 –z 1.07
```

We will end up with something like what's shown in the following screenshot:

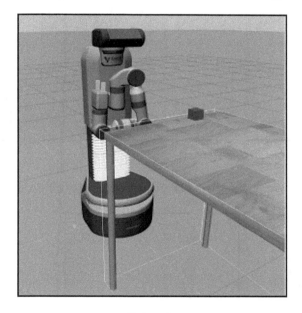

Simulator robot

Next, we will launch MoveIt with perception:

```
$ roslaunch myrobot_moveit_config myrobot_planning_execution.launch
```

The output is as follows:

Visualization in RViz

Like we did previously, we will set the **Planning Attempts** option to something like 7:

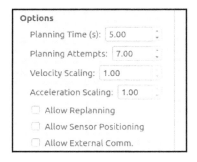

Planning Attempts

Next, we will launch the grasping server we created in the previous experiment:

```
$ roslaunch myrobot_grasping grasp_generator_server.launch
```

Finally, we will launch this Python script and view the planning of robot and its action in Rviz:

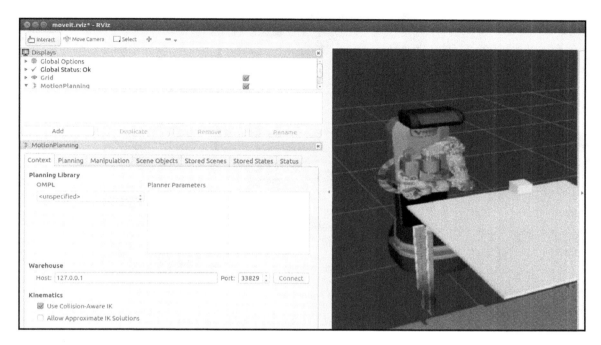

RViz view

The preceding screenshot shows the perception and motion planning views of the robot in RViz. Correspondingly, the following diagram shows the motion of the robot in the simulation environment:

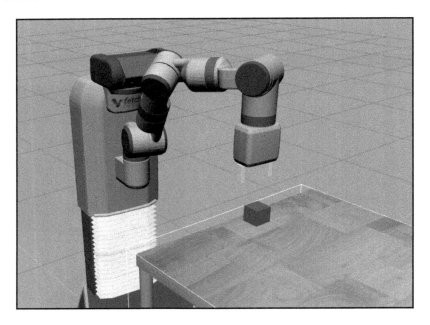

Simulator view

And that's all! We've finished this chapter! Well... almost finished.

Micro Aerial Vehicles in ROS

In this chapter, we will discuss the following recipes:

- Overview of MAV system design
- A generic mathematical model of an MAV/drone
- Simulation of an MAV/drone using RotorS/Gazebo
- Autonomous navigation framework for an MAV/drone
- Working with a real MAV/drone – Parrot, Bebop

Introduction

In this chapter, we will learn about a modular **micro aerial vehicle** (**MAV**) simulator framework, which enables us to quick start performing research and development on MAVs. After completing this chapter, we will have a ready-to-use MAV simulator, including control and state estimation. The simulator will be designed in a modular way, where different controllers and state estimators can be implied interchangeably to incorporate any new MAVs in a few steps by only changing a parameter file. Moreover, we will also compare different controllers and state estimators with the provided evaluation framework.

The simulator framework is a good starting point for handling higher-level tasks such as collision avoidance and path planning with **Simultaneous Localization and Mapping** (**SLAM**). We will discuss the design of all components, which will be analogous to their real world counterparts. Thus, this enables the usage of the same controllers and state estimators, including their configuration parameters, in the simulation as well as in the real MAV.

Overview of MAV system design

The research and development of algorithms on MAVs needs access to expensive hardware, and field experiments usually consume a considerable amount of time, as well as requiring a trained pilot for execution. However, most of the errors, occurring on real platforms, are tough to reproduce, which usually results in damaging the vehicle. Hence, the RotorS simulator framework has been developed to reduce field experiment times, which makes debugging easier and finally reduces the crashes of real MAVs. Moreover, this is also useful for student projects, where access to an expensive real platform may not be permitted.

To perform high-level experiments such as collision avoidance and path planning, including SLAM for autonomous navigation, the simulator framework could be used for the provided model of MAVs. In addition, the framework also includes a position controller and a state estimator.

Getting ready

In this chapter, we will describe in detail the steps required to set up the RotorS simulator framework, shown in the following screenshot, including the **Robot Operating System (ROS)** and Gazebo. Once this chapter has been completed, we will be able to set up the simulator, attach basic sensors to a MAV, and make it able to navigate autonomously in the virtual world:

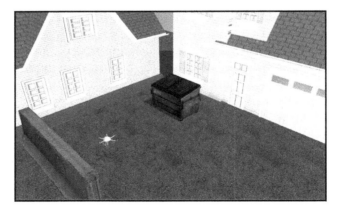

RotorS simulator

We will also be able to compare the algorithms using the evaluation scripts. Finally, we will discuss how all the aspects learned and methods developed in this chapter could be then applied to a real MAV.

An overview of the primary components of the RotorS simulator is shown in the following diagram. Although, in this chapter, we will focus on the simulation parts, which are shown on the left side in the following diagram, a lot of effort has been made to keep the structure of the simulator similar to the real system. Ideally, all high-level components used in a simulated environment could be run on the real platform without any major changes:

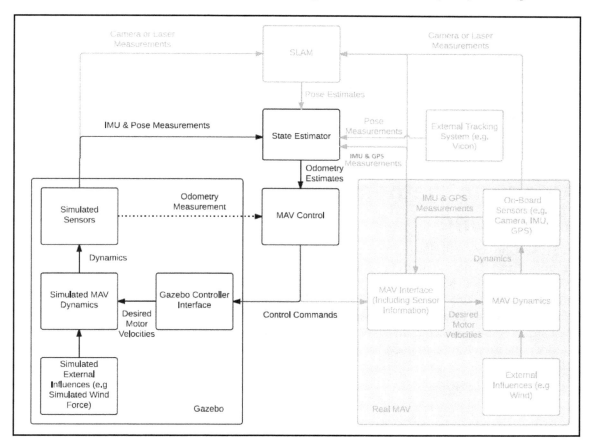

MAV system design

All components, used on real MAVs, can be simulated by Gazebo plugins and a physics engine. We have developed the simulator as a modular way of assembling MAVs, where an MAV consists of a body frame, a fixed number of rotors, which can be placed at specified locations, and several sensors, which can be attached to the body as required. Each rotor has its own motor dynamics, the parameters of which can be identified on a real MAV, a **Firefly** from Ascending Technology using recorded flight data. Similarly, several sensors, such as an **Inertial Measurement Unit** (**IMU**), a stereo camera, GPS, and sensors developed by the user can be attached to the body frame. Moreover, we will also implement noise models for the applied sensors to simulate realistic conditions.

We will discuss an implementation of a geometric controller with a simple interface to facilitate the development of different control strategies. This provides access to various levels of commands, such as angular rates, attitude, or position control.

One of the important components is the state estimation, which is used to obtain information about the state of the MAV at a high rate. Although state estimation is crucial on real MAVs, in the simulation state parameters such as position, orientation, linear, and the angular velocity of the MAV are directly provided by a Gazebo plugin.

A generic mathematical model of an MAV/drone

In this section, we will discuss the mathematical concepts needed to understand the kinematic and dynamic model of MAVs that is used in simulator design. Throughout this chapter, we refer to MAVs as multirotor quadcopters, but the concepts can be applied to other vehicles to. One important concept will be the kinematics and dynamics of an MAV, which allow us to develop control strategies on different command levels such as attitude or position and orientation. In addition, we will also discuss the state estimation of a quadcopter, which will be used for higher-level tasks such as path planning or local collision avoidance.

Getting ready

A quadcopter is a multirotor helicopter that is lifted and propelled by four rotors, as shown in the following photograph. So, what are the important components and their functions? A quadcopter consists of two pairs of identical propellers, where one pair of propellers rotates in a clockwise direction and the other pair rotates in an counter-clockwise direction. These propellers are connected to motors, which are driven by an electronics speed controller. A quadcopter uses independent variations of speed from each rotor to achieve control. By chaining the speed of each rotor, it is possible to specially generate the desired total *thrust* and desired total *torque*.

A quadcopter also consists of a microcontroller; a microcontroller is just a small computer based on embedded devices such as Arduino. This microcontroller is connected to different sensors, such as a speedometer and a gyroscope, which provide state information—position, velocity, acceleration, and orientation. An electronic receiver is connected to the microcontroller, which transfers instructions from the remote control to the microcontroller. All these devices are powered by the battery:

Quadcopter

Forces and moments on quadcopter

Four propellers produce *thrust* F_i in a direction perpendicular to the plane of rotation of propellers, as shown in the following diagram. This *thrust* is proportional to the square of the angular velocity of the propeller:

$$F_i = K_f \times \omega_i^2$$

$$M_i = K_m \times \omega_i^2$$

$$M_y = (F_{1-} F_2) \times L$$

$$M_y = (F_{3-} F_4) \times L$$

$$Weight = mg$$

These forces are shown in action in the following diagram:

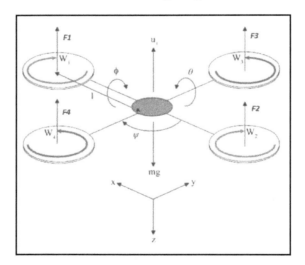

Forces and moments on a quadcopter

As the propellers rotate, they create the reaction *Moments* M_i on the quadcopter, around the z axis . This reaction *moments* is proportional to the square of the angular velocity of the propeller. The thrust produced by the opposite pair of propellers produces *moments* M_x and M_y respectively. These *moments* are given by a difference of forces multiplied by the *length* between the two propellers. Last but not the least, there is the force of gravity, which is always acting in a downward direction.

The motion of the quadcopter can be analyzed by Newton's Second Law of Motion:

For linear motion:

$Force = mass \times linear\ acceleration$

For angular/rotation motion:

$Torque = Intertia \times angular\ acceleration$

Hover and upward and downward motion

Let us look at some conditions of the motion of the quadcopter; first, how the quadcopter can hover in the air! As the quadcopter is steady in the air, it must be in equilibrium. All the forces must be balanced. The total *thrusts* F_1, F_2, F_3, and F_4 produced by the propellers must be equal to the weight of the quacopter. And, all the *Moments* produced must be zero:

Condition for hovering:

$$mg = F_1 + F_2 + F_3 + F_4$$

$All\ Moments = 0$

Equation of motion:

$$m\ddot{r} = F_1 + F_2 + F_3 + F_4 - mg$$

$$m\ddot{r} = 0$$

The governing equation motion of in this condition is $m\ddot{r} = 0$.

In the case of upward motion, the net *Thrust* produced by all the propellers will be more than the weight of the quadcopter. This result represents an upward motion of the quadcopter; in such cases, the $m\ddot{r} > 0$ must hold.

The condition for upward motion is:

$$mg < F_1 + F_2 + F_3 + F_4$$

$All\ Moments \neq 0$

Equation of motion:

$$m\ddot{r} = F_1 + F_2 + F_3 + F_4 - mg$$

$$m\ddot{r} > 0$$

For downward motion, the *thrust* produced by the propeller must be less than the weight of the quadcopter.

Condition for downward motion:

$$mg > F_1 + F_2 + F_3 + F_4$$

$$All\ Moments \neq 0$$

The equation of motion is:

$$m\ddot{r} = F_1 + F_2 + F_3 + F_4 - mg$$

$$m\ddot{r} < 0$$

Rotation (yaw) motion

The yaw motion is the rotation of the quadcopter in the horizontal plane. As stated earlier, two propellers rotate in a clockwise direction and the other two propellers rotate in a counter-clockwise direction. These pairs of propellers produce the reaction *Moments* in opposite directions. If the reaction *Moments* produced by each of propeller pairs is equal and opposite, then there is no yaw motion. If the rotation *Moment* produced by one pair of propellers is more than the *Moment* produced by the other pair of propellers, then there is a net resultant *Moment* produced, which causes the rotation of the quadcopter about its vertical axis.

The condition for rotation motion is:

$$mg = F_1 + F_2 + F_3 + F_4$$

$$All\ Moments \neq 0$$

The equation of motion is:

$$m\ddot{r} = F_1 + F_2 + F_3 + F_4 - mg$$

$$I_{zz} \ddot{\Psi} = M_1 + M_2 + M_3 + M_4$$

Linear (pitch and roll) motion

The pitch and roll motion is the rotation of the quadcopter about its horizontal axis. In this case, the opposite pairs of propellers produce unequal forces, which cause net non-zero *moments*.

Condition for linear motion:

$$mg < F_1 + F_2 + F_3 + F_4$$

$$All\ Moments \neq 0$$

Equation of motion:

$$m\ddot{r} = F_1 + F_2 + F_3 + F_4 - mg$$

$$I_{xx} \times \ddot{\Phi} = (F_3 - F_4) \times L$$

$$I_{yy} \times \ddot{\Theta} = (F_1 - F_2) \times L$$

These cause the rotation of the quadcopter about its horizontal axis. For linear motion in the horizontal plane, the pitch (Φ) and roll (Θ) angles must be non-zero. This causes non-zero components of Thrust in a horizontal direction. This force causes resultant *Moments* in the horizontal plane.

Simulation of an MAV/drone using RotorS/Gazebo

In this section, we will discuss how to use the RotorS simulator. We will also discuss an overview of the different components of the simulator. Then, we will explain how to use our controllers to get the MAV into hovering mode and perform linear and rotation motion. We will also explain how sensors can be added, and how to use our evaluation scripts for validation of state estimation. In addition, we will learn advanced topics such as how to build a custom model, controller, or sensor plugin.

Getting ready

Before installing the MAV simulator RotorS, the package manager should be used to install the necessary dependencies on Ubuntu 16.04 with ROS Kinetic:

```
$ sudo sh -c 'echo "deb http://packages.ros.org/ros/ubuntu `lsb_release -
sc` main" > /etc/apt/sources.list.d/ros-latest.list'
$ wget http://packages.ros.org/ros.key -O - | sudo apt-key add -
$ sudo apt-get update
$ sudo apt-get install ros-kinetic-desktop-full ros-kinetic-joy ros-
kinetic-octomap-ros ros-kinetic-mavlink python-wstool python-catkin-tools
protobuf-compiler libgoogle-glog-dev ros-kinetic-control-toolbox
$ sudo rosdep init
$ rosdep update
$ source /opt/ros/kinetic/setup.bash
```

The source installation of RotorS is independent of the operating system and ROS version, so we recommend the same. First of all, we will make sure that the catkin workspace is located at `~/catkin_ws/src` and get the simulator and additional dependencies as follows:

```
$ cd ~/catkin_ws/src
$ git clone https://github.com/ethz-asl/rotors_simulator.git
$ git clone https://github.com/ethz-asl/mav_comm.git
$ cd ~/catkin_ws /
$ catkin build
$ source ~/catkin_ws/devel/setup.bash
```

We would also like to use the demos with an external state estimator, so the following additional packages are needed:

```
$ cd ~/catkin_ws/src
$ git clone https://github.com/ethz-asl/ethzasl_msf.git
$ git clone https://github.com/ethz-asl/rotors_simulator_demos.git
$ git clone https://github.com/ethz-asl/glog_catkin.git
$ git clone https://github.com/catkin/catkin_simple.git
$ cd ~/catkin_ws /
$ catkin build
$ source ~/catkin_ws/devel/setup.bash
```

Simulator overview

We can split up the RotorS simulator into various packages, as shown in the following diagram. We can incorporate an already existing robot into the simulation by describing its geometry and kinematic properties:

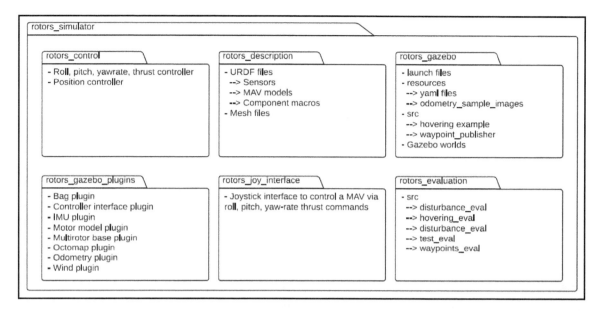

A modular Gazebo MAV simulator framework

A simulator design can be divided into the following steps:

1. Model selection, you can either:
   - Pick one of the models provided by RotorS
   - Build a custom model

2. Attaching sensors to the MAV, you can either:
   - Use one of the Xacro for the sensors shipped with RotorS
   - Develop a custom sensor and attach it directly

3. Adding a controller to the MAV, you can either:
   - Start one of the controllers shipped with RotorS
   - Develop a custom controller

4. A state estimator, you can do one of the following:
   - Use the output of the ideal odometry sensor
   - Use MSF, described in the comming section
   - Develop a custom state estimator

How to do it...

To check if the setup is working, we want to start with a short example. We will validate our setup by running a simulator of an AscTec Firefly hex-rotor helicopter, which has an inbuilt position controller. Here, we have an ideal sensor that publishes odometry data that can directly be used by a controller.

Hovering

The following command will start the simulation:

```
$ roslaunch rotors_gazebo mav_hovering_example.launch
```

In the following screenshot we can see the hex-rotor helicopter taking off after 5 seconds, and flying to the point *P = (0,0,1)*:

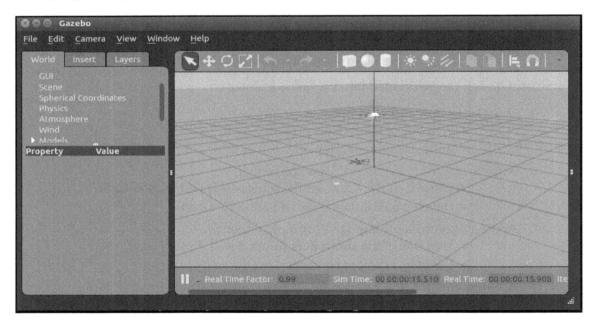

Hovering

In addition, the following diagram shows the ROS communication network, which gives an overview of all the ROS nodes that are running, and the topics on which the nodes are communicating:

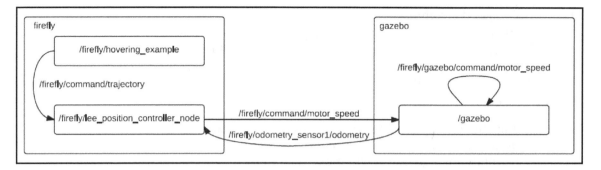

Hovering – ROS communication network

Although, in the preceding screenshot, Gazebo is only shown as one ROS node, internally all the Gazebo plugins are running, such as IMU and individual motors that are mounted on the frame. For example, a generic odometry sensor is mounted on the Firefly, which publishes the odometry messages in the /firefly/odometry_sensor1 namespace, as shown in the following listing:

```
kbipin@kumar:~/catkin_ws$ rostopic list
/clock
/firefly/command/motor_speed
/firefly/command/pose
/firefly/command/trajectory
/firefly/gazebo/command/motor_speed
/firefly/ground_truth/imu
/firefly/ground_truth/odometry
/firefly/ground_truth/pose
/firefly/ground_truth/pose_with_covariance
/firefly/ground_truth/position
/firefly/ground_truth/transform
/firefly/imu
/firefly/joint_states
/firefly/motor_speed
/firefly/motor_speed/0
/firefly/motor_speed/1
/firefly/motor_speed/2
/firefly/motor_speed/3
/firefly/motor_speed/4
/firefly/motor_speed/5
/firefly/odometry_sensor1/odometry
/firefly/odometry_sensor1/pose
/firefly/odometry_sensor1/pose_with_covariance
/firefly/odometry_sensor1/position
/firefly/odometry_sensor1/transform
/firefly/wind_speed
```

Firefly message topics

The odometry message includes the position, orientation, and the linear and angular velocity of the MAV. Hence, the position controller can directly subscribe to the odometry message. Furthermore, the controller publishes actuator messages, which are read by the Gazebo controller interface and forwarded to the individual motor model plugins.

The controller get the commands from `MultiDOFJointTrajectory` messages, which are published by the `hovering_example` node. These messages contain references of poses and its derivatives, however, in this example only the position and the values of the messages are set. Even if the `MultiDOFJointTrajectory` messages are compatible with planners, such as MoveIt! But, waypoints can also be published as `PoseStamped` messages in some cases.

We can change the position reference of the MAV by sending a `MultiDOFJointTrajectory` message using a planner or `waypoint_publisher`. For example, the following command will move the Firefly to the given pose (x, y, z, yaw):

```
$ rosrun rotors_gazebo waypoint_publisher 5 0 1 0 __ns:=firefly
```

The output can be seen in the following screenshot:

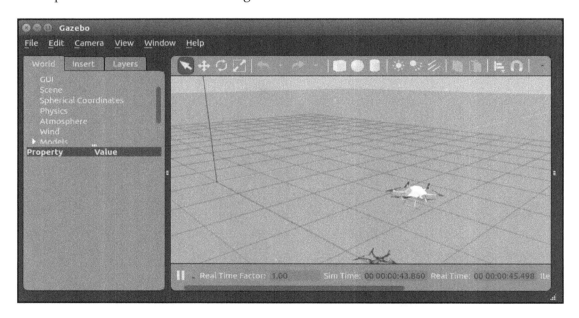

Movement for Firefly

In addition, we can exchange the vehicle by setting the `mav_name` argument as illustrated in the following command:

```
$ roslaunch rotors_gazebo mav_hovering_example.launch mav_name:=pelican
```

The output can be seen in the following screenshot:

Pelican

In current RotorS packages, `asymmetric_quadrotor`, `firefly`, `hummingbird`, and `pelican` MAVs are available:

- All the control parameters are set in `lee_controller_<mav_name>.yaml`, and the vehicle parameters used by the controller in `<mav_name>.yaml`, which are not identical
- On real systems, the vehicle parameters are usually unknown, and only approximate values can be estimated

State estimation

However, in the real MAVs, generally, there is no direct odometry sensor, such as has been used in the previous subsection. Rather, there is a broad variety of sensors, such as GPS and magnetometer, cameras, or lasers to do SLAM, or external tracking systems such as motion capture setup that provide a full **six degrees of freedom** (**6DoF**) for pose. In this section, we will discuss how to use MSF packages to get the full state from a pose sensor and the IMU.

We will run the example that needs the `rotors_simulator_demos` package that was downloaded in the previous subsection. The following command will start all the required ROS nodes:

```
$ roslaunch rotors_simulator_demos mav_hovering_example_msf.launch
```

We can observe, in the following screenshot, that this time the MAV is not as stable as in the previous example, which had a small offset at the beginning. Here, wobbling comes from the simulated noise on the pose sensor and the offset from the IMU biases. Nevertheless, after a while the offset will disappear, because the **Extended Kalman Filter (EKF)** estimates the biases correctly:

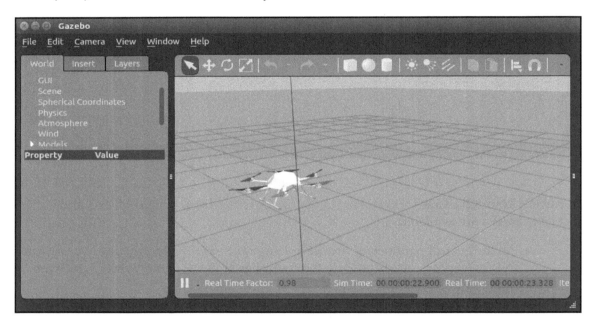

MSF demo

Here, looking at the `re-drawn rqt_graph` in the following diagram, we can see that an additional node is started. The MSF `pose_sensor` and the controller node now subscribe to the odometry topic from the MSF, instead of the odometry from Gazebo:

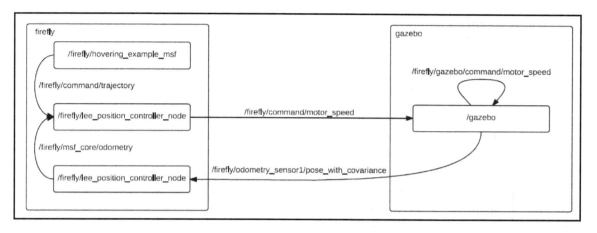

MSF – ROS communication network

As discussed in the previous subsection, the following command can be used to move the MAV:

```
$ rosrun rotors_gazebo waypoint_publisher 5 0 1 0 __ns:=firefly
```

The MSF package performs the state estimation based on IMU measurements, which can use any predefined available sensor combinations, and we use `pose_sensor` in this example, where the parameters are loaded from the `msf_simulator.yaml` located in the `rotors_simulator_demos/resources` folder.

Sensors mounting

In this section, we will discuss the different sensors and how these are mounted on the MAVs. Any sensor can be attached to the vehicle in the corresponding Xacro file. In our example, the `mav_hoverin_example.launch` file loads the robot description from the `firefly_generic_odometry_sensor.gazebo` file, which adds an odometry sensor to the Firefly, by calling the following:

```
<xacro:odometry_plugin_macro (here go the macro properties )>
  <inertia (with i t s properties ) />
  <origin (with i t s properties ) />
</ xacro:odometry_plugin_macro>
```

The macro properties are explained in the `component_snippets.xacro` file. Properties present in all of RotorS sensor macros are:

- `namespace`: ROS namespace assigned to given sensor
- `parent_link`: The sensor gets attached to this link
- `some_topic`: The name of the topic on which the sensor publishes messages

An overview of the sensors currently used in RotorS, for which Xacros are provided in *Table 1*:

Sensor	Description	Xacro-name Gazebo plugins
Camera	A standard ROS camera	`camera_macro` `libgazebo_ros_camera.so`
IMU	An IMU implementation, with zero mean white Gaussian noise	`imu_plugin_macro` `librotors_gazebo_imu_plugin.so`
Odometry	Simulated odometry sensors	`odometry_plugin_macro` `librotors_gazebo_odometry_plugin.so`

VI-Sensor	This is a stereo camera with an IMU, featuring embedded synchronization and time-stamping developed at Autonomous Systems Lab (ASL)	`vi_sensor_macro`
		`libgazebo_ros_camera.so`
		`librotors_gazebo_odometry_plugin.so`
		`librotors_gazebo_imu_plugin.so`
		`libgazebo_ros_openni_kinetic.so`

Tabel 1: An overview of the Xacros sensors provided in the RotorS simulator

The Firefly hexacopter has a VI-sensor mounted on it, pointing to the front and slightly downwards, as shown in the following photograph. For more information about the VI-sensor visit `http://www.skybotix.com/`:

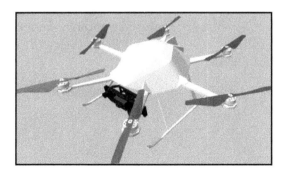

Firefly with VI-sensor

Evaluation

We are especially interested in tracking the MAV's state in a body fixed frame with respect to a user-specified inertial frame of reference, and how it reacts to disturbances. Moreover, a great interest lies also in observing how quickly an MAV can fly to a certain location, and again how accurately it stays around the specified setpoint.

In the first experiment, we will create a bag file of a flight with a Firefly, running the hovering example with logging enabled, by using the following command:

```
$ roslaunch rotors_gazebo mav_hovering_example.launch enable_logging:=true
```

The bag files will be stored in `.ros-folder` in the user's `Home` directory by default. Next, we will evaluate the controller by executing the following scripts:

```
$ rosrun rotors_evaluation hovering_eval.py -b
~/.ros/firefly_2018-04-06-11-32-58.bag --save_plots True --mav_name firefly
```

Sometimes, there is an error about an unindexed bag file. We can reindex it with:

```
$rosbag reindex <bagfile>
```

We will get the following output from the script:

```
Position RMS error : 0.050 m
Angular velocity RMS error : 0.001 rad/s
```

Additionally, the script will produce three plots showing the position, the position error, and the angular velocities. The following graphs shows the position error and the angular velocities respectively:

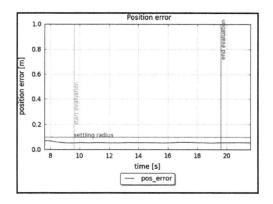

Position error

The preceding graph shows the position error, and the following graph shows the angular velocity with respect to the time:

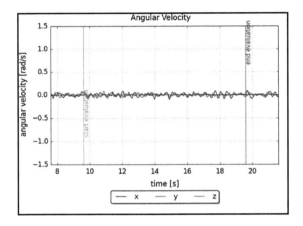

Angular velocity

We can achieve more realistic conditions by adding some noise on the odometry sensor. The noise parameters of the odometry sensor can be set in the `firefly_generic_odometry_sensor.gazebo` file. There are evaluation scripts available for multiple waypoints `waypoint_eval.py` and for evaluating the reaction to external disturbances `disturbance_eval.py`.

How it works...

Until now, we have learned how to use the the components provided by RotorS. If this set of MAVs with some standard sensors and controllers cannot fulfill our requirement, in this section we will give more insight on how to develop a custom controller. Here, we will discuss how to integrate a new MAV, how to write new sensors, and how to work on a state estimator.

Developing a custom controller

We will discuss how to build a controller for an arbitrary MAV. The message passing is managed by `gazebo_controller_interface`, and the motor dynamics are handled in `gazebo_motor_model`. Thus, the task of developing a custom controller can be reduced to subscribing to state estimator messages, reference commands, and publishing actuator messages on the `command/motor_speed topic`. Alternatively, we can also execute the `RollPitchYawrate-ThrustController` and develop a custom position controller that publishes `MultiDOFJointTrajectory` messages. Here, the custom position controller subscribes to `MultiDOFJointTrajectory` messages as a control input, whereas the `roll-pitch-yawrate-thrust` controller subscribes to `RollPitchYaw-rateThrust` messages.

We can split the design of the controller into two parts; the first part handles the the parameters and the message passing, and the second part of the controller is a library, which does all the computations. Moreover, we recommend using the template of controller ROS-nodes, located in the `rotors_control/src/nodes` folder. It reads the controller parameters from the `ROS` parameter server as YAML files, which get split up into controller specific parameters and vehicle specific parameters. The vehicle parameter file includes the mass of the MAV, its inertia, and rotor configuration, whereas controller gains are specified in the controller parameters file.

Our custom controller libraries are located in the `rotors_control/src/library` folder. Another time, we recommend using one of the libraries as a template to develop our custom controller. We have a method, `CalculateRotorVelocities`, that gets called in every control loop, and computes the required rotor velocities ω, based on the state information of the MAV.

There's more...

We are able to fly with the MAVs that are shipped with RotorS. But it would be great if we were able to bring our own MAV into RotorS. In this section, we will discuss how to design and integrate our custom MAV into RotorS. In addition, we describe how custom sensors can be written as a Gazebo plugin.

We will use the **Unified Robot Description Format** (**URDF**), with **XML Macros** (**Xacro**) for the description of the robot, which was discussed in detail in `Chapter 6`, *Robot Modeling and Simulation*.

 Although, Gazebo has its own format to describe the robots, objects, and the environment, called **Simulation Description Format** (**SDF**), we are adhering firmly to URDF here, as this format can be displayed in the RViz. Nevertheless, all the SDF-specific properties can be added by putting them in a `<gazebo>` block. Internally, Gazebo converts the URDF files to SDF files.

If we have a specific robot that we would like to use in our simulation, the procedure is quite straightforward. First of all, we have to identify which parts of our robot are fixed rigid bodies; in URDF these parts are called links and the connections of these links are called joints.

In the following diagram, the different links and joints of a asymmetric quadrotor helicopter are shown, where each joint has a parent and a child link. There are a number of different joints, as can be seen at `http://wiki.ros.org/urdf/XML/joint`, however for our application we only use three types: continuous joints, fixed joints, and revolute joints:

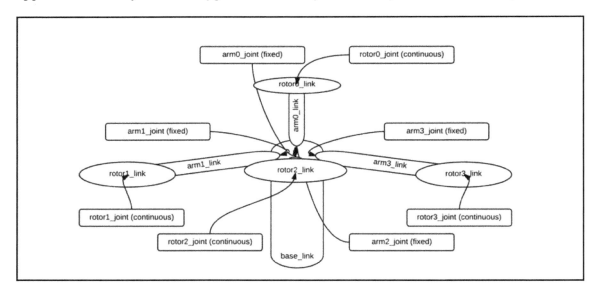

A design draft of a multirotor helicopter with four non-symmetrical aligned rotors

As we go on with this section, we will learn how to develop a model of a conceptual quadcopter or quadrotor helicopter, as depicted in the preceding diagram. This quadrotor helicopter model has asymmetric design that has four rotors, of which three are distributed equally on the edge of a circle with radius *l*, while the remaining rotor is placed in the back of the vehicle with an arm of length *l/3*. However, all the rotors are of the same dimension.

Now, we can describe our model in a URDF file. The robot, which is described using URDF, needs a `<robot>` tag in the beginning of the file, with a unique name; within the opening- and the closing tag, the robot will be assembled. We can start by adding the base of the robot as follows:

```
<robot name="asymmetric_quadrotor" xmlns:xacro="http://ros.org/wiki/xacro">
  <xacro:multirotor_base_macro
    robot_namespace="${namespace}"
    mass="${mass}"
    body_width="${body_width}"
    body_height="${body_height}"
    use_mesh_file="${use_mesh_file}"
    mesh_file="">
    <xacro:insert_block name="body_inertia"/>
  </xacro:multirotor_base_macro>
</robot>
```

The preceding code listing needs some Xacro-properties, such as namespace and mass, to be set. We can continue adding the arms and the rotors to the base link by connecting them with joints—we will discuss this for one arm and rotor as an example.

As discuss in `Chapter 6`, *Robot Modeling and Simulation*, if we have a mesh file of our multirotor helicopter with arms, and know its inertia, we can skip this step and attach the rotors directly to the base link. Nevertheless, in all of the other MAVs that are provided with RotorS, the rotors are directly attached to the base. We would like to encourage the use of the macros provided in the `component_snippets.xacro` and `multirotor_base.xacro` files:

```
<link name="arm1_link">
  <xacro:box_inertial x="${arm_length}" y="0.03" z="0.01"
mass_box="${mass_arm}" />
    <visual>
      <origin xyz="${arm_length/2} 0 0" rpy="0 0 0" />
      <geometry>
        <box size="${arm_length} 0.03 0.01" />
      </geometry>
    </visual>
    <collision>
      <origin xyz="${arm_length/2} 0 0" rpy="0 0 0" />
```

```
        <geometry>
          <box size="${arm_length} 0.03 0.01" />
        </geometry>
      </collision>
  </link>
  <joint name="arm1_joint" type="fixed">
    <origin xyz="0 0 0" rpy="0 0 ${2*pi/3}" />
    <parent link="base_link" />
    <child link="arm1_link" />
  </joint>
  <xacro:vertical_rotor
    robot_namespace="${namespace}"
    suffix="1"
    direction="cw"
    motor_constant="${motor_constant}"
    moment_constant="${moment_constant}"
    parent="arm1_link"
    mass_rotor="${mass_rotor}"
    radius_rotor="${radius_rotor}"
    time_constant_up="${time_constant_up}"
    time_constant_down="${time_constant_down}"
    max_rot_velocity="${max_rot_velocity}"
    motor_number="1"
    rotor_drag_coefficient="${rotor_drag_coefficient}"
    rolling_moment_coefficient="${rolling_moment_coefficient}"
    color="Blue"
    use_own_mesh=" false "
    mesh="">
  <origin xyz="${arm_length} 0 ${rotor_offset_top}" rpy="0 0 0" />
  <xacro:insert_block name=" rotor_inertia "/>
  </xacro:vertical_rotor>
```

Since we are using a very similar design snippet for all four arms, it is a good idea to develop a macro, which can be named as `arm_with_rotor`. This will take the arm length, an angle, the motor number, and the mass as parameters.

We can check our design description by using the following command:

```
$ rosrun xacro xacro .py <xacro_file> > <urdf_file>
$ check_urdf <urdf_file>
```

Finally, we can spawn our asymmetric quadrotor helicopter in an empty Gazebo world:

```
$ roslaunch rotors_gazebo mav_hovering_example.launch
mav_name:=asymmetric_quadrotor
```

We will see an asymmetric quadrotor, as in the following screenshot:

Asymmetric quadcopter

See also

To make our robot model perceive the environment and plan its own motion, we have to add sensors to our model. We have already discussed how to add available sensors in a previous subsection. Thus, we will look into how to develop a Gazebo plugin for a new sensor.

Creating custom sensors

We create a plugin for a sensor only when no sensor plugin is available that satisfies our requirements. In this subsection, we will describe how one would conceptually proceed with developing a wind sensor. A wind sensor usually measures the airspeed v_{air} on a robot. This airspeed is the difference between the wind speed v_{wind} wind and the current velocity v of the robot:

$$v_{air} = v_{wind} - v$$

To integrate this sensor into the simulation, we can follow the procedure for creating a Model-Plugin, as described at `http://gazebosim.org/tutorials?tut=plugins_model&cat=write_plugin`.

We will create a subscriber to the wind topic, and store the pointer to the link, which the plugin gets attached to. Additionally, we will create a publisher on the desired topic, for example `air_speed`. All of this can be implemented in the `Load` method of the plugins. In the `OnUpdate` method, which gets called at every single simulation iteration, we are getting the link's velocity in world frame by calling `GetWorldLinearVel()`:

```
link_- >GetWorldLinearVel()
```

And perform the calculation described previously, in the equation. We can also add some additional noise to the the calculated value. Moreover, we can refer the `GazeboImuPlugin` and the `GazeboOdometryPlugin` to get an idea of how to add noise to our calculated value.

Autonomous navigation framework for an MAV/Drone

One of the greatest advantages of the RotorS simulator is that it comes with a fully integrated and functional trajectory tracking controller. This motivates us to implement higher-level tasks such as collision avoidance and path planning without having to implement state estimation and the controller first. Additionally, in simulation, we also have access to perfect ground truth, which is usually hard to get on real systems.

Getting ready

Overall, we have the required setup and resources to experiment and develop a comprehensive autonomous navigation framework for any MAV. Once we have a working solution in a simulation, transitioning to real world systems is typically an engineering task. In this section, we would like to discuss some ideas for how to implement collision avoidance and path planning in the RotorS simulator.

Collision avoidance

A commonly used approach to solve collision avoidance on MAVs is to project the environment onto a 2D ground plane and configure the ROS navigation stack, as discussed in `Chapter 7`, *Mobile Robot in ROS*. To handle this problem, an appropriate sensor needs to be mounted on the MAV, to perceive the environment and to get an estimate of the surroundings. Fortunately, Gazebo provides plugins for 2D lasers such as Hokuyo that could likely be mounted on a real MAV. However, this approach cannot be substantial in the real world since the operating spaces of the MAV are reduced to a plane at a constant height.

The solution for full 3D collision avoidance requires front looking depth cameras such as the Kinect-sensor, which is already implemented in Gazebo. This could be a good starting point. These types of sensors are lightweight and provide rich information about the surroundings.

Alternatively, the monocular camera can be mounted on the MAV, thereby implementing structure from motion, which will provide a 3D estimate of the environment.

The following photograph shows the working of **Parallel Tracking and Mapping** (**PTAM**) (`http://www.robots.ox.ac.uk/~gk/PTAM/`) on a Parrot AR drone:

PTAM with Parrot AR Drone

Similarly, a stereo camera can be used to calculate the disparity images and perform 3D collision avoidance directly on the disparity images.

Moreover, we can also use ORB-SLAM (http://webdiis.unizar.es/~raulmur/orbslam/) for state estimation and 3D mapping for environments using a monocular fish-eye camera mounted on a Parrot Bebop drone, which is shown in the following screenshot:

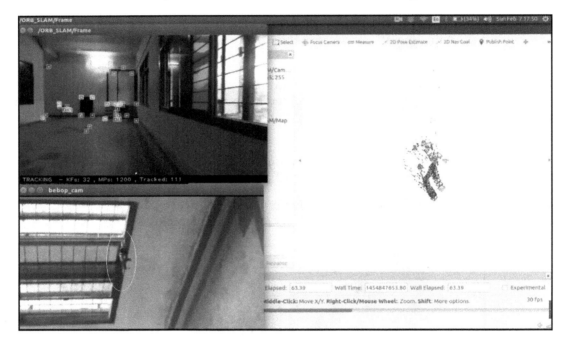

ORB-SLAM with Parrot Bebop Drone

Path planning

In the RotorS simulator, we can prepare a demo environment with a power plant and a waypoint publisher to experiment with different planning algorithms. A file with the Octree representation of a demo environment including the power plant can be found in the `rotors_gazebo/resources` folder and can be used with the `octomap_server` package. The Octree representation allows for efficient collision checking in the planner and is applicable for 3D environments.

The following command will execute our experimental setup:

```
$ roslaunch rotors_gazebo mav_powerplant_with_waypoint_publisher.launch
```

Following is the output:

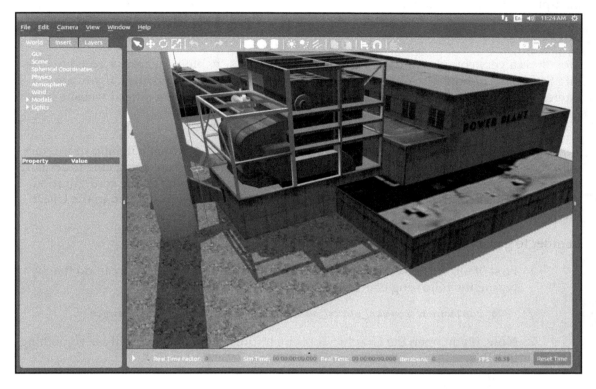

MAV: Path planning in a power plant

ROS also provides a large collection of planning algorithms in the MoveIt! package, which was discussed in Chapter 8, *Robotic Arm in ROS*. These can also be configured to perform 3D path planning in a static environment and control the MAV.

How to do it...

In this section, we are going to see how we can perform drone navigation using MoveIt!

MoveIt! is an ROS framework that allows us to perform motion planning with a specific robot. And what does this mean? Well, it basically means that it allows us to plan movements (motions) from point A to point B, without colliding with anything.

An important thing to mention is that MoveIt! is, by default, prepared to work with manipulation robots, and that's what it's usually used for. Therefore, we will have to take some extra steps in order to use it for our purpose, which is to plan trajectories for our drone.

MoveIt! is a very complex and useful tool. So, within this chapter, we are not going to dive into the details of how MoveIt! works, or all of the features that it provides. If you are interested in learning more about MoveIt!, you can have a look at the official website here:

```
http://moveit.ros.org/
```

Fortunately, MoveIt! provides a very nice and easy-to-use GUI, which will help us interact with the robot in order to perform motion planning. However, before being able to actually use MoveIt!, we need to build a package. This package will generate all the configuration and launch files required for using our defined robot (the one that is defined in the URDF file) with MoveIt!

In order to generate this package, we will just follow these steps:

1. First of all, we'll need to launch the **MoveIt Setup Assistant**. We can do that by typing the following command:

```
$ roslaunch moveit_setup_assistant setup_assistant.launch
```

2. Now, if you open the Graphic Tools by hitting **Start** icon, you will see something like this:

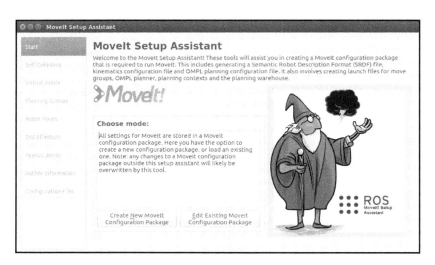

MoveIt Setup Assistance

Great! We are now in the **MoveIt Setup Assistant**. The next thing we'll need to do is load our robot file. So, let's continue.

3. Click on the **Create New MoveIt Configuration Package** button. A new section will appear, as shown in the following screenshot:

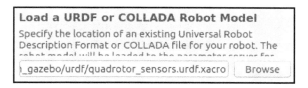

Load new MoveIt Configuration Package

4. Now, we will just click the **Browse** button, select the Xacro file named `quadrotor_sensors.urdf.xacro` located in the `chapter9_tutorials/model` package, and click on the **Load Files** button. We should now see something similar to the following screenshot:

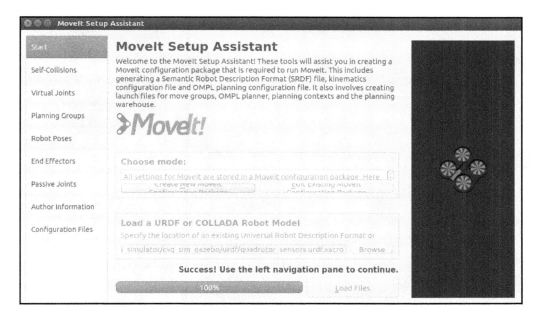

MoveIt! and quadcopter URDF model

Great! So, we've loaded our robot's Xacro file to the **MoveIt Setup Assistant**. Now, let's start configuring some things.

5. Next, we will go to the **Self-Collisions** tab, and click on the **Regenerate Default Collision Matrix** button. We will end up with something similar to the following screenshot:

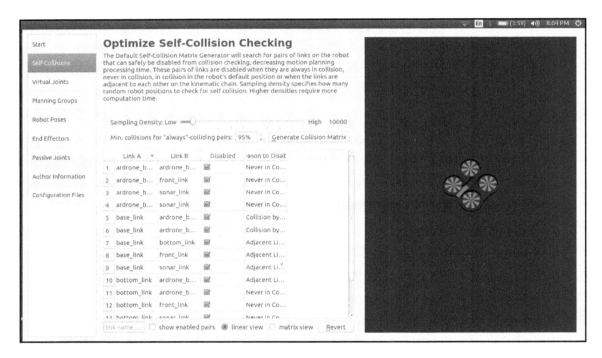

Self-Collision Checking

Here, we are just defining pairs of links that don't need to be considered when performing collision checking. For instance, because they are adjacent links, they will always be in collision.

6. Next, we will move to the **Virtual Joints** tab. Here, we will define a virtual joint for the base of the robot by clicking the **Add Virtual Joint** button, and set the name of this joint to `virtual_joint`. The **Child Link** will be `base_link`, and the **Parent Frame Name** will be `world`. We will also set the **Joint Type** to `floating`. This is all shown in the following screenshot:

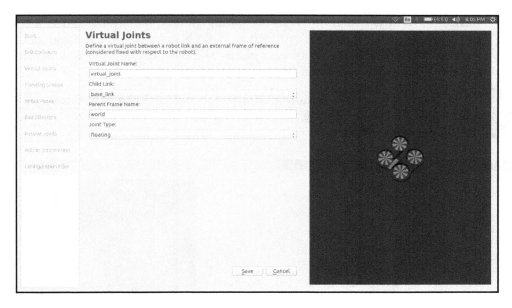

Virtual Joints

7. Finally, we will click the **Save** button:

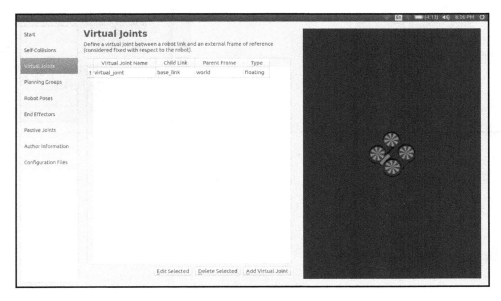

Virtual Joints List

Basically, what we are doing here is creating an imaginary joint that will connect the base of our robot to the simulated world. This virtual joint represents the motion of the base of the robot on a plane. This virtual joint will be a floating joint since the quadrotor is a multi-degree-of-freedom object.

8. Now, we will open the **Planning Groups** tab and click the **Add Group** button. Here, we will create a new group called `ardrone_group`, as shown in the following screenshot:

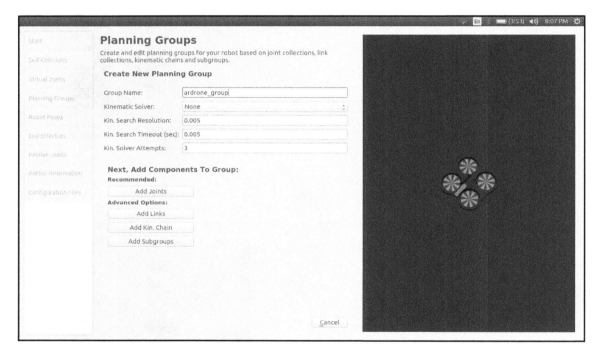

Planning Groups

9. Next, we will click on the **Add Joints** button, and we will add the **virtual_joint**:

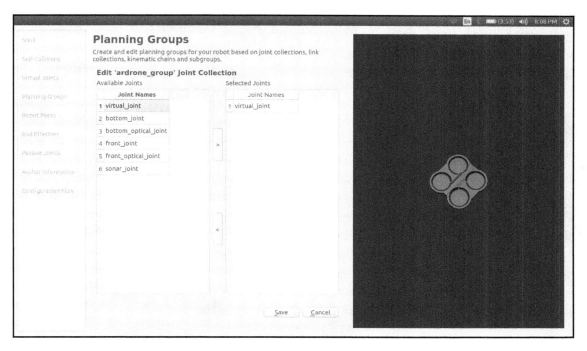

Add Joints

10. Now, we will repeat the same process, except for adding a link. In this case, we will add the **base_link**, as shown in the following screenshot:

Saving the Joints' configuration

11. Finally, we will click the **Save** button, and we will end up with something like this:

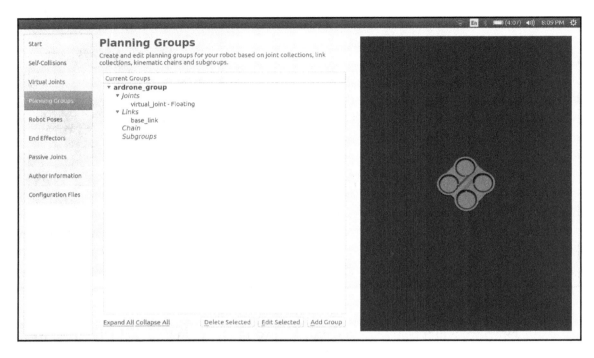

View the Planning Groups

Planning Groups are used for semantically describing different parts of the quadrotor, such as defining an end effector. For our quadrotor model, we don't have an end effector, but we define a planning group named `ardrone_group`, which contains the floating `virtual_joint` and the `base_link`. We can leave the kinematic solver set to none since we are treating the quadrotor as a simple single object.

12. Next, we can skip directly to the **Author Information** tab. Here, we will just enter our name and email.

13. Finally, we will go to the **Configuration Files** tab, click the **Browse** button, and navigate to the `catkin_ws/src` directory, where we will create a new directory, and name it `ardrone_moveit_config`. Now, we will choose the just created directory, as shown in the following screenshot:

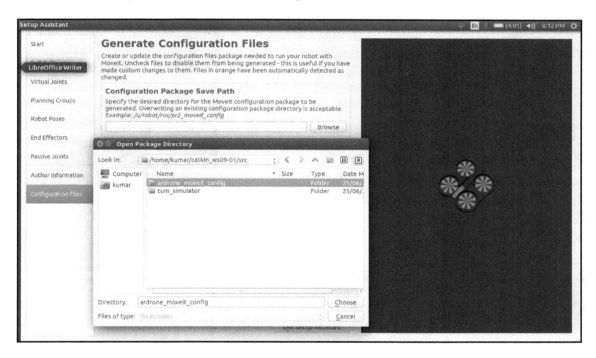

Generate Configuration Files

14. Now, we will click the **Generate Package** button. If everything goes well, we should see something similar to the following screenshot:

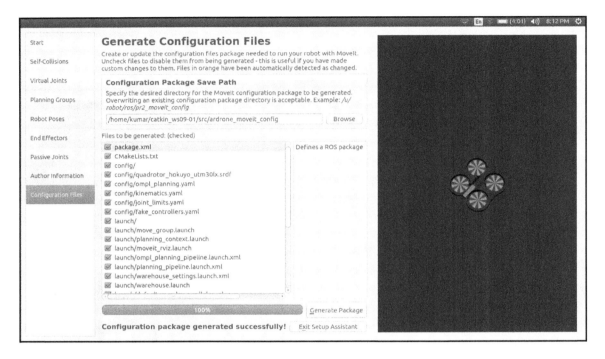

Configuration completed

And that's it! We've created our MoveIt! package for our drone. But... now what?

15. Well… to start, we can launch the MoveIt! RVIZ environment and start doing some tests in order to check that everything is working fine. So, we will follow the next exercise to do so:

```
$ roslaunch ardrone_moveit_config demo.launch
```

16. If everything goes OK, we will see something similar to the following screenshot:

MoveIt launch with quadcopter

17. Now, we will move to the **Planning** tab, as shown in the following screenshot:

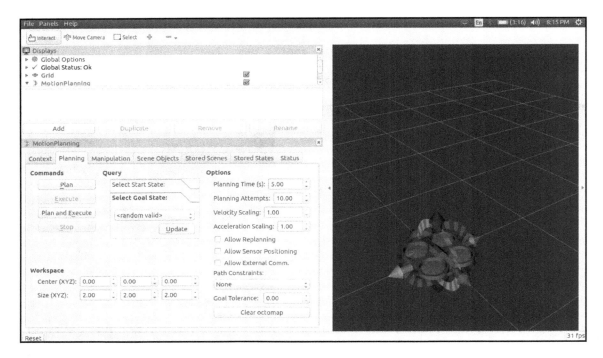

Planning tab

18. Before we start planning anything, it is always a good practice to update the **Start State** to **<current>**:

State selection

19. Now, we will have to set the workspace to a value bigger than the current one. Remember that MoveIt! is, by default, a tool for manipulator robots, which usually operate in small workspaces. In our case, though, we are going to work with a drone, so we will need to make our workspace bigger. For instance, we can set it to 10 x 10 x 10, as shown in the following :

Workspace

Center (XYZ):	0.00	0.00	0.00
Size (XYZ):	10.00	10.00	10.00

Workspace setting

20. We can move the drone using the interactive arrows. In this manner, move it to a desired goal position and click on the **Plan** button. We will see how a trajectory is planned, as shown in the following screenshot:

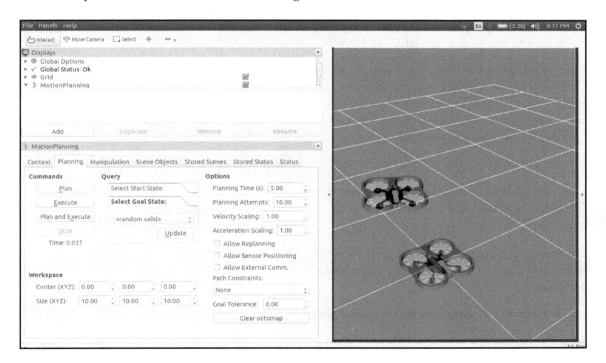

Planning trajectory

Now, we have just played with the new tool! We can repeat this same process a few more times. Also, we can try to play with the workspace values, or turn the different displays on and off to see how they affect the simulation.

So, we've already checked that everything works as expected, and we're also a little more familiar with MoveIt! The next step will be to integrate the environment into MoveIt!, because we need to be aware of which environment we want our drone to navigate in if we want to avoid obstacles. We are going to do this by using an OctoMap.

An OctoMap is, basically, a 3D occupancy grid map of an environment. The map implementation is based on an Octree. OctoMaps are provided by the OctoMap library. We can find more information regarding the OctoMap library at http://octomap.github.io/.

OctoMap is incorporated into MoveIt! as a plugin (called the Occupany Map Updator plugin), which can update Octree from different kinds of sensor inputs, such as PointClouds and depth images from 3D vision sensors.

In our case, though, we do not have any 3D sensors mounted on our drone, so we won't be able to use this plugin. Instead, we are going to do a simple and fast workaround in order to be able to introduce an OctoMap into the MoveIt! planning scene.

First of all, since we don't have any 3D sensors on our drone, we will have to build this OctoMap in some other way. In this case, we have built the OctoMap using a Husky robot, which does have a Kinect camera mounted on it. OctoMaps can be built using the octomap_mapping ROS package. We can find more information on how to build an OctoMap with the following link:

http://wiki.ros.org/octomap_mapping

Once the OctoMap is created, we can easily save it into a file to use later. In our case, this OctoMap is stored in a file called simple_octomap.bt, located in the chapter9_tutorials/maps package.

Once we have created and saved an OctoMap, we can provide this OctoMap using the octomap_server ROS package. In order to provide an OctoMap, we can use the following command:

```
$ rosrun octomap_server octomap_server_node /path_to_octomap_file
```

In our case, the command will be as follows:

```
$ roscd chapter9_tutorials;
$ cd maps
$ rosrun octomap_server octomap_server_node small_octomap.bt
```

Great! But now… how do we know that my OctoMap is being published somewhere? Well, let's find out!

Once we have executed the preceding command, the full OctoMap contained within the specified file is published into a topic named `/octomap_full`, as a `octomap_msgs/Octomap` message. Also, the OctoMap is published into the topic `occupied_cells_vis_array` as box markers for visualization in RViz, using the `visualization_msgs/MarkerArray` message.

So, if we want to visualize the OctoMap in RViz, we will have to open a RViz window, add a **Marker Array** element, and configure it with the `occupied_cells_vis_array` topic. We should then be able to visualize something like this:

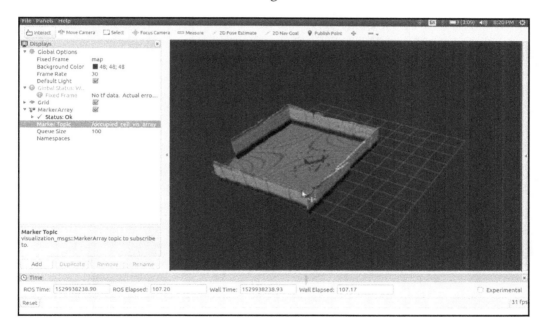

OctoMap

Excellent! So, now we've seen that our OctoMap is being properly published. But how can we add this OctoMap to the MoveIt! planning scene? Let's see!

First of all, we will create a new package that will be named `load_octomap` and add `rospy` as a dependency:

```
$ catkin_create_pkg load_octomap rospy
```

Inside the `src` folder of this new package, create a new file named `octoload.py` and copy the following content into it:

```python
#! /usr/bin/env python

import rospy
from octomap_msgs.msg import Octomap
from moveit_msgs.msg import PlanningScene, PlanningSceneWorld

class OctoHandler():
    mapMsg = None

    def __init__(self):
        rospy.init_node('moveit_octomap_handler')
        rospy.Subscriber('octomap_full', Octomap, self.cb, queue_size=1)
        pub = rospy.Publisher('move_group/monitored_planning_scene',
PlanningScene, queue_size=1)
        r = rospy.Rate(0.25)
        while(not rospy.is_shutdown()):
            if(self.mapMsg is not None):
                pub.publish(self.mapMsg)
            else:
                pass
            r.sleep()

    def cb(self, msg):
        psw = PlanningSceneWorld()
        psw.octomap.header.stamp = rospy.Time.now()
        psw.octomap.header.frame_id = 'map'
        psw.octomap.octomap = msg

        psw.octomap.origin.position.x = 0
        psw.octomap.origin.orientation.w = 1

        ps = PlanningScene()
        ps.world = psw
        ps.is_diff = True
        self.mapMsg = ps
if __name__ == '__main__':
    octomap_object = OctoHandler()
```

Don't worry if you don't understand much of this code right now; we are going to have a look at it later.

Inside our package, we will create a new folder named `launch`, and add a `launch` file there to start this code. We can name it `load_octomap.launch`. Also, add a `<node>` tag for providing the OctoMap. It can be something like the following:

```
<launch>

    <arg name="path"
default="/home/user/simulation_ws/src/small_octomap.bt"/>

    <node pkg="octomap_server" type="octomap_server_node"
name="octomap_talker" output="screen" args="$(arg path)">
    </node>
    <node pkg="load_octomap" type="octoload.py"
name="moveit_octomap_handler" output="screen">
    </node>
</launch>
```

Next, we will execute our `launch` file, and launch the MoveIt! environment again, as we have done before in this chapter. We can check that the OctoMap is properly loaded into the MoveIt! planning scene, as shown in the following screenshot:

MoveIt with OctoMap

Great! So, we are now able to load our OctoMap into the MoveIt! planning scene. But how did we achieve this? What did the code that was used really do? Let's have a look:

```
from octomap_msgs.msg import Octomap
from moveit_msgs.msg import PlanningScene, PlanningSceneWorld
```

Here, we are importing some messages that we need. The OctoMap message should be able to handle the OctoMap, and the `PlanningScene` and `PlanningSceneWorld`, should be able to publish into the MoveIt! planning scene:

```
class OctoHandler():
    mapMsg = None
```

Here, we are just creating a class called `OctoHandler()`, and initializing a `mapMsg` variable:

```
rospy.init_node('moveit_octomap_handler')
        rospy.Subscriber('octomap_full', Octomap, self.cb, queue_size=1)
        pub = rospy.Publisher('move_group/monitored_planning_scene',
PlanningScene, queue_size=1)
        r = rospy.Rate(0.25)
        while(not rospy.is_shutdown()):
            if(self.mapMsg is not None):
                pub.publish(self.mapMsg)
            else:
                pass
            r.sleep()
```

This is the constructor of the class. Here, we are basically defining a `Subscriber` of the `/octomap_full` topic so that we are able to get the OctoMap from this topic. Remember, we have seen earlier that this is the topic where the OctoMap is published when we provide it through the `small_octomap.bt` file. Also, we define a `Publisher` that will allow us to publish a message into the `move_group/monitored_planning_scene`, which is the topic that MoveIt! uses to build the planning scene.

Finally, we create a loop for publishing our OctoMap into the MoveIt! planning scene. The `mapMsg` always has a value (this value will be the OctoMap):

```
def cb(self, msg):
        psw = PlanningSceneWorld()
        psw.octomap.header.stamp = rospy.Time.now()
        psw.octomap.header.frame_id = 'map'
        psw.octomap.octomap = msg

        psw.octomap.origin.position.x = 0
        psw.octomap.origin.orientation.w = 1
```

```
ps = PlanningScene()
ps.world = psw
ps.is_diff = True
self.mapMsg = ps
```

This is the callback for the `Subscriber` we defined before. Every time a message is published into the `/octomap_full` topic, this callback will be activated. Inside this callback, we are basically building the `PlanningScene()` message, which will be the one that we will use to publish into the `move_group/monitored_planning_scene` topic.

First, we build a `PlanningSceneWorld()` message, then we fill it with the OctoMap message, along with other required values:

```
psw = PlanningSceneWorld()
    psw.octomap.header.stamp = rospy.Time.now()
    psw.octomap.header.frame_id = 'map'
    psw.octomap.octomap = msg

    psw.octomap.origin.position.x = 0
    psw.octomap.origin.orientation.w = 1
```

We then pass this message to the `world` variable of the `PlanningScene()` message:

```
ps = PlanningScene()
    ps.world = psw
    ps.is_diff = True
```

Finally, we save this message into the `mapMsg` variable of the class, which is the variable we are going to publish into the `move_group/monitored_planning_scene` topic. Remember the loop in the previous snippet of code:

```
if(self.mapMsg is not None):
                pub.publish(self.mapMsg)
```

And that's it! Now everything makes more sense, right?

So, at this point, we have our MoveIt! package built, and we can publish our OctoMap into the MoveIt! planning scene so that the drone is aware of its environment. This way, it can avoid the obstacles that are in its way. But we still have a couple of things to do. First of all, we need to connect the MoveIt! environment to the real drone (in this case, for obvious reasons, the real drone is the simulated one). Up until this point, we have only been working with the drone within the MoveIt! application. This is very useful because you can do many tests without worrying about any damage. But the final goal will always be to move the real robot, right?

How it works...

The MoveIt! package we've created is able to provide the necessary ROS services and actions in order to plan trajectories, but it isn't able to pass these trajectories to the real robot to execute. All the plans we've done were executed in an internal simulator that MoveIt! provides. In order to communicate with the real drone, it will be necessary to do some modifications to the MoveIt! package we created at the beginning of this section.

But before that, we need to learn a couple of things about how MoveIt! works.

The primary node used in MoveIt! is the `move_group` node. As shown in the system architecture in the following diagram, the `move_group` node integrates all of the external nodes in order to provide a set of ROS actions and services for users. These external nodes include the quadrotor sensors and controllers, data from the parameter server, and a user interface. We can read more about MoveIt! concepts here:

http://moveit.ros.org/documentation/concepts/

The system architecture appears as follows:

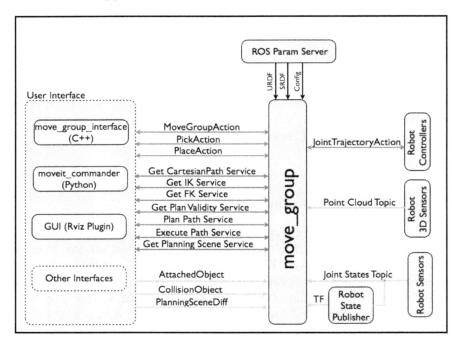

MoveIt system architecture

The move_group node communicates with the drone using ROS topics and actions. The node receives the current state information, such as the position and orientation of the drone, by listening to the /joint_states topic. Therefore, it will be necessary to launch a joint_state_publisher node to broadcast the state of the drone. The move_group node also receives global information about the quadrotor's pose using the ROS TF library. The TF provides the transformation between the base frame of the robot and the world or map frame. In order to publish this information, a robot_state_publisher node is executed.

The move_group node interacts with the quadrotor's control system through the FollowJointTrajectoryAction interface. But, as we have already mentioned at the beginning of the section, the MoveIt! package was originally created with the manipulation of robot arms in mind, which is not what we are doing here. Therefore, the FollowJointTrajectoryAction action will have to be modified to accommodate the multi-DOF dynamics of the drone. This new action will be called MultiDofFollowJointTrajectory. In this system, this action is already provided, so we don't have to bother creating it. We just have to configure MoveIt! to use it. If we want to check out more information about it, we can have a look at the following repository, developed by Alessio Tonioni:

https://github.com/AlessioTonioni/Autonomous-Flight-ROS

Great! So… with all this new knowledge, let's begin doing some proper modifications in order to connect the MoveIt! package to the real drone.

First of all, we will have to create a file to define how we will control the joints of our real robot. Inside the config folder of our MoveIt! package, we will create a new file named controllers.yaml and copy the following content into it:

```
controller_list:
  - name: multi_dof_joint_trajectory_action
    type: MultiDofFollowJointTrajectory
    default: true
    joints:
      - virtual_joint
```

So, basically, here we are defining the action server that will be used for controlling the joints of our robot. In this case, the only joint we have on our drone is the virtual_joint. And the action server we are going to use in order to control it is the MultiDofFollowJointTrajectory, as explained previously. The name parameter defines the name of the controller, and the default parameter specifies if the controller is the primary controller chosen by MoveIt! for communicating with a particular set of joints.

Next, we'll have to create a file to define the limits of the joints of the drone. For our drone, we are going to limit its maximum velocity and acceleration. Again, inside the `config` folder, create a new file called `joint_limits.yaml`, and copy the following content into it:

```
joint_limits:
  virtual_joint:
    has_velocity_limits: true
    max_velocity: 0.2
    has_acceleration_limits: true
    max_acceleration: 0.04
```

Now, if we open `smart_grasping_sandbox_moveit_controller_manager.launch.xml`, which is inside the `launch` directory, we will find it empty and so copy the following content into it:

```
<launch>
 <!-- Set the param that trajectory_execution_manager needs to find the
controller plugin -->
 <arg name="moveit_controller_manager"
default="moveit_simple_controller_manager/MoveItSimpleControllerManager" />
 <param name="moveit_controller_manager" value="$(arg
moveit_controller_manager)"/>
 <!-- load controller_list -->
 <rosparam file="$(find ardrone_moveit_config)/config/controllers.yaml"/>
</launch>
```

What we are doing here is basically loading the `controllers.yaml` file we have just created, and the `MoveItSimpleControllerManager` plugin, which will allow us to send the plans calculated in MoveIt! to our real robot—in this case, the simulated drone.

Finally, we will have to create a new launch file that sets up the entire system to control our robot. So, inside the `launch` folder, create a new launch file called `ardrone_navigation.launch`:

```
<launch>
  <param name="use_sim_time" value="true" />
  <!-- Take the name of the package in which config is stored-->
  <arg name="config_pkg" default="$(find ardrone_moveit_config)" />

  <!-- By default, we are not in debug mode -->
  <arg name="debug" default="false" />

  <!-- Load the URDF, SRDF and other .yaml configuration files on the param
server -->
```

```xml
  <include file="$(arg config_pkg)/launch/planning_context.launch">
    <arg name="load_robot_description" value="true"/>
  </include>
  <node pkg="tf" type="static_transform_publisher"
name="world_to_footprint" args="0 0 0 0 0 0 world odom 5" />
  <node pkg="tf" type="static_transform_publisher" name="odom_to_nav"
args="0 0 0 0 0 0 odom nav 5" />
  <node pkg="tf" type="static_transform_publisher"
name="virtual_joint_broadcaster_0" args="0 0 0 0 0 0 base_footprint
base_link 5" />
  <node pkg="tf" type="static_transform_publisher"
name="odom_map_broadcaster" args="0 0 0 0 0 0 map world 5" />

  <!-- We do not have a robot connected, so publish fake joint states -->
  <node name="joint_state_publisher" pkg="joint_state_publisher"
type="joint_state_publisher">
    <param name="/use_gui" value="false"/>
    <rosparam
param="/source_list">[/move_group/fake_controller_joint_states]</rosparam>
  </node>

  <!-- Given the published joint states, publish tf for the robot links -->
 <node name="robot_state_publisher" pkg="robot_state_publisher"
type="robot_state_publisher" respawn="true" output="screen" />

 <node name="action_controller" pkg="action_controller"
type="action_controller" ></node>

  <!-- Run the main MoveIt executable without trajectory execution (we do
not have controllers configured by default) -->
  <include file="$(arg config_pkg)/launch/move_group.launch">
    <arg name="allow_trajectory_execution" value="true"/>
    <arg name="fake_execution" value="true"/>
    <arg name="info" value="true"/>
    <arg name="debug" value="$(arg debug)"/>
  </include>

  <!-- Run Rviz and load the default config to see the state of the
move_group node -->
  <include file="$(arg config_pkg)/launch/moveit_rviz.launch">
    <arg name="config" value="true"/>
    <arg name="debug" value="$(arg debug)"/>
  </include>
</launch>
```

Finally, we are starting the launch files that we need in order to set up the MoveIt! environment. The most important things here are the following:

```
<node pkg="tf" type="static_transform_publisher" name="world_to_footprint"
args="0 0 0 0 0 0 world odom 5" />
  <node pkg="tf" type="static_transform_publisher" name="odom_to_nav"
args="0 0 0 0 0 0 odom nav 5" />
  <node pkg="tf" type="static_transform_publisher"
name="virtual_joint_broadcaster_0" args="0 0 0 0 0 0 base_footprint
base_link 5" />
  <node pkg="tf" type="static_transform_publisher"
name="odom_map_broadcaster" args="0 0 0 0 0 0 map world 5" />
```

Here, we are publishing some transforms that we need, since we are publishing our OctoMap in an unusual way. Also, we are connecting some links that are not connected by default. This is because, as we've been saying throughout the chapter, MoveIt! is not a tool that was initially created to navigate drones:

```
<node name="action_controller" pkg="action_controller"
type="action_controller" ></node>
```

We are starting the `MultiDofFollowJointTrajectory` action server that we talked about in the beginning of this section:

```
<include file="$(arg config_pkg)/launch/move_group.launch">
    <arg name="allow_trajectory_execution" value="true"/>
    <arg name="fake_execution" value="true"/>
    <arg name="info" value="true"/>
    <arg name="debug" value="$(arg debug)"/>
</include>
```

This is the main MoveIt! launch, which launches all of the MoveIt! environments.

Great! We have finished all of the MoveIt! setup in order to be able to calculate trajectories for our drone within an environment that is represented by the OctoMap. But let's do a very quick summary in order to organize all the recent information that we've received.

Typically, MoveIt! relies on predefined action files, such as `FollowJointTrajectory` and `GripperCommand` actions. Since MoveIt! wasn't developed with a drone's navigation in mind, we have defined our own action. The `MultiDofFollowJointTrajectory` action defines the goal, feedback, and result fields needed to enable MoveIt! to develop and transmit multi-DOF trajectories for the drone to follow. This action is already built into the system, so we don't have to worry about it.

The quadrotor is treated as a single multi-DOF joint. Thus, we defined the `controllers.yaml` file, in order to tell MoveIt! to use this action. This action generates trajectories in the form of a set of waypoints that the drone has to follow.

Now what we'll have to do is add some kind of code that is capable of reading these waypoints and transforming them into real movements by the drone.

We will discuss executing the trajectory with the real drone in next section, after getting familiar with ROS packages available for most popular real drones, such as Parrot and Bebop.

Working with a real MAV/drone – Parrot, Bebop

Although it is nice, of course, to have a simulation of an MAV such as RotorS, it always raises the question of how well it represents the real world scenario and how feasible the transition from simulation to real MAVs would be. However, during the design and development of RotorS simulator, a lot of effort has been invested into keeping the model of the simulator as close as possible to the real system.

Getting ready

In the best case, in an ideal world, just switching the simulation environment with an ROS node that communicates with the hardware will get real MAVs to perform high-level tasks such as autonomous navigation, including collision avoidance and path planning.

The simulator can be a great tool that enables the development of algorithms, to be deployed on a real MAV later. Then, the transition of the code from the simulator to the code running on the actual hardware will be as simple as possible, but only if the interface is designed in a way that mimics most of the interfaces on the real systems. Additionally, it is also necessary to incorporate real world uncertainty using a probabilistic approach. One of the major challenges on real platforms is managing the timing delays that result in a non-deterministic order of measurements.

How to do it...

To have confidence and give an impression of how well the simulator replicates the real MAV, we are using the same controller gains in the simulation as on the real Hummingbird and Firefly MAVs.

The ROS provide the `ardrone_autonomy` and `bebop_autonomy` packages for the Parrot AR Drone and the Parrot Bebop Drone respectively, as shown in the following photograph:

Parrot AR and Bebop Drones

Following are the packages:

- `ardrone_autonomy`: https://ardrone-autonomy.readthedocs.io/en/latest/
- `bebop_autonomy`: https://bebop-autonomy.readthedocs.io/en/latest/

Similarly, the AscTec Pelican/ Hummingbird and Firefly quadrotor helicopters are built by Ascending Technologies and can be interfaced with ROS using an `asctec_autopilot` package. These helicopters are shown in the following photograph:

AscTec Pelican/Hummingbird and Firefly

The `asctec_autopilot` package is available at `http://wiki.ros.org/Robots/AscTec`.

Executing the trajectory with the real MAV/drone

MoveIt! does not currently come with the ability to execute the multi-DOF trajectories required to take full advantage of the quadcopter's dynamics. Instead, we will have to add some custom code in order to handle the trajectories generated by the `MultiDofFollowJointTrajectoryAction` action server.

For that, we will use a couple of Python scripts that have already been developed by the eYSIP. We can find their repository at `https://github.com/eYSIP-2017`.

First of all, we will download the `tum_ardrone` package, using the following command in the `/src` folder of our workspace:

```
$ git clone https://github.com/tum-vision/tum_ardrone.git
```

Next, let's compile the package:

 ROS packages for Ardrone and Debop are available for ROS Indigo, it requires effort to make it work on kinetic and later distributions of ROS.

The first step will be to publish the OctoMap, as we learned in the previous section:

```
$ roslaunch load_octomap load_octomap.launch
```

Next, let's start the RViz environment by executing the launch file we created in the previous section:

```
$ roslaunch ardrone_moveit_config ardrone_navigation.launch
```

Let's start the waypoint action server by issuing the following command:

```
$ rosrun drone_application move_to_waypoint.py _real_drone:=false
_aruco_mapping:=false
```

And now, let's have the drone take off:

```
$ rostopic pub /ardrone/takeoff std_msgs/Empty [TAB][TAB]
```

We will see how the drone takes off in MoveIt!

Next, let's go to the MoveIt! window, select a goal position for the drone, using the interactive arrows, and plan a trajectory to get there.

Finally, execute the follow trajectory node in order to execute the planned trajectory:

```
$ rosrun drone_application follow_trajectory.py _real_drone:=false
_aruco_coords:=false _visualise_trajectory:=false
```

How it works...

Let's do a quick summary of the code to better understand what's going on:

Basically, we have two main files, `move_to_waypoint.py` and `follow_trajectory.py`.

The functionality of the `move_to_waypoint.py` script is to create an action server that is able to convert the waypoints-based trajectory into real movement of the drone. So it handles the actual movement of the drone and publishes the feedback:

```
class moveAction(object):
```

The action server is defined inside a class named `moveAction`:

```
def __init__(self, name, real_drone, aruco_mapping):
```

The `init()` function is the constructor of the class. It creates the simple action server, alongside other things that are necessary to move the drone. For instance, it also creates a Publisher to send messages to the `/cmd_vel` topic.

We can see both of these in the lines as follows:

```
# Creates the Publisher
self.pub = rospy.Publisher('/cmd_vel', Twist, queue_size=5)
# Creates the Action Server
self._as = actionlib.SimpleActionServer(
            self._action_name,
            drone_application.msg.moveAction,
            execute_cb=self.execute_cb,
            auto_start=False)
```

The `moniter_transform()` function provides data about the current pose of the drone based on the odometry readings:

```
def moniter_transform(self):
```

The `get_camera_pose()` function is only used when working with ArUco markers, which is not the case here. So, we are not using this function:

```
def get_camera_pose(self, temp_pose):
```

The `move_to_waypoint()` function is the most important one here. This is the function that actually makes the drone move. It takes the waypoint and the current pose of controls PID. Then, it calls the PID controller until the current pose is equal to the waypoint received:

```
def move_to_waypoint(self, waypoint):
```

Finally, the `execute_cb()` function is the callback function of the action server. Each time a new goal is sent to the action server, this `callback` function is activated. It basically calls the `move_to_waypoint()` function, passing a waypoint:

```
def execute_cb(self, goal):
```

The functionality of the `follow_trajectory.py` script is to create an action client that extracts the waypoints contained in the trajectory, which are generated by MoveIt!, and send them as goals to the `moveAction` action server created in the `move_to_waypoint.py` script.

This `send_trajectory()` function, as its name says, sends the waypoints to be followed to the action server. It sends each individual waypoint as a goal to the `moveAction` action server:

```
def send_trajectory(waypoints, client=None):
```

This `get_waypoints()` function, basically, extracts the waypoints from the MoveIt-generated trajectory (which is published in the `move_group/display_planned_path` topic). This function has a very dirty implementation:

```
def get_waypoints(data):
```

```
def legacy_get_waypoints(data):
```

The `legacy_get_waypoints()` function has the same utility as the immediately preceding function, but it has a much cleaner implementation. This is the one that is actually used.

As we can see at the end of this `legacy_get_waypoints()` function:

```
def legacy_get_waypoints(data):
```

Right after sending the trajectory to be executed, it publishes a message into the `/ardrone/land` topic:

```
# once waypoints are ready send to move_to_waypoint
send_trajectory(waypoints) # land once the trajectory is executed
land_pub.publish()
```

This is why, after executing the trajectory, the drone lands. If we want, we can try to change this behavior by modifying a little bit of the code.

And that's it! We finally know the whole process to navigate a drone using MoveIt!

ROS-Industrial (ROS-I) **10**

In this chapter, we will discuss the following recipes:

- Understanding ROS-I packages
- 3D modeling and simulation of an industrial robot and MoveIt!
- Working with ROS-I packages – Universal Robots, ABB robot
- ROS-I robot support packages
- ROS-I robot client package
- ROS-I robot driver specification
- Developing a custom MoveIt! IKFast plugin
- ROS-I-MTConnect integration
- Future of ROS-I – Hardware support, capabilities, and applications

Introduction

ROS-I is a project whose main goal is to bring ROS closer to the robotics industrial world. At this point, ROS is already a massive tool for roboticists all around the world, but it is mostly used by research centers and universities, or by service robot companies. The industrial robotics world is a difficult area to access openly, mainly because they usually work with closed-loop motions, based on tasks that are very repetitive and mechanized. Lately, the industrial world has become more interested in performing more complex and dynamic tasks, making their robots more intelligent.

As a consequence, this is where ROS-I plays its role! This is because the ROS-I package comes with a solution to interfacing and controlling industrial robot manipulators with ROS, using its powerful tools, such as MoveIt!, Gazebo, and `rviz`. It is important to note that it is an open-source project, so anyone can develop the applications for different robots, which will, in the end, benefit the industrial sector.

Understanding ROS-I packages

ROS-I enhances the advanced capabilities of ROS software for industrial robots which are employed in manufacturing processes. ROS-I packages are BSD (legacy)/Apache 2.0 (preferred) licensed programs, which contain libraries, drivers, and tools with a standard solution for industrial hardware. ROS-I is now guided by the ROS-I Consortium. The official website of ROS-I can be found at `http://rosindustrial.org/`.

The following diagram shows the official logo of ROS-I Consortium (ROS-I):

ROS-I logo

In 2012, the ROS-I open source project with the collaboration of *Yaskawa Motoman Robotics* (`http://www.motoman.com/`), *Willow Garage* (`https://www.willowgarage.com/`), and the *Southwest Research Institute (SwRI)* (`http://www.swri.org/`). The ROS-I was founded by Shaun Edwards in January 2012.

Later, in March 2013, the ROS-I Consortium Americas and ROS-I Consortium Europe were launched by *SwRI* and the *Fraunhofer Institute for Manufacturing Engineering and Automation (Fraunhofer IPA)*, respectively.

The objective behind ROS-I development can be characterized as follows:

- Bringing together the strength of ROS with the existing industrial technologies to develop a reliable and robust software for industrial robot applications
- Providing a platform for research and development in industrial robotics which could build a wide community supported by researchers and professionals for industrial robotics altogether
- Developing ROS-I, open source software that allows commercial use without any restrictions

Getting ready

ROS-I packages can be installed by using package managers or can be built from the source code. Although we have installed the `ros-kinetic-desktop-full` in the past, we can use the following command to install ROS-I packages on Ubuntu:

```
$ sudo apt-get install ros-kinetic-industrial-core
```

Moreover, we can find the complete repository of ROS-I at `https://github.com/ros-industrial` and download the required packages. The preceding command will install the core packages of ROS-I, which consist of the following set of ROS packages:

- `industrial-core`
- `industrial_deprecated`
- `industrial_msgs`
- `simple_message`
- `industrial_robot_client`
- `industrial_robot_simulator`
- `industrial_trajectory_filters`

We could discuss each of the layers in brief for better understanding, but the architecture block diagram is available at the ROS-I Wikipedia page (`http://wiki.ros.org/Industrial`), which is shown in the following diagram:

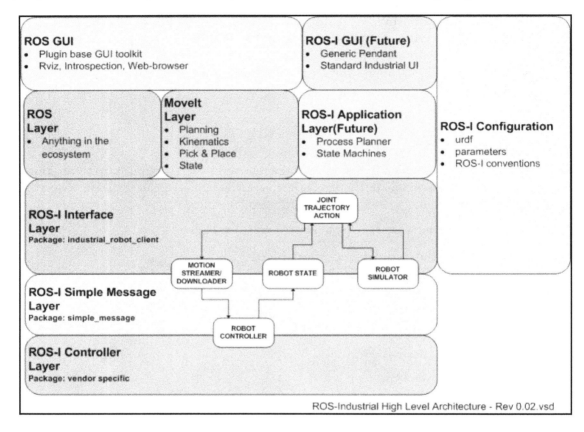

ROS-I architecture block diagram

Let's understand the components in detail:

- **ROS GUI**: This layer consists of the ROS plugin-based GUI tools, such as `rviz`, `rqt_gui`, and so on
- **ROS-I GUI**: These are standard UIs for working with industrial robots that will be implemented in the future
- **ROS Layer**: This is the base layer for the ROS Middleware Framework, where all communications are taking place

- **MoveIt!Layer**: The MoveIt! layer provides a solution to industrial manipulators in planning, control, and execution
- **ROS-I Application Layer**: This layer consists of an industrial process planner for manufacturing
- **ROS-I Interface Layer**: This layer consists of the industrial robot client, which can be interfaced to the industrial robot controller using the simple message protocol
- **ROS-I Simple Message Layer**: This layer consists of a standard set of protocols that will send data from the robot client to the controller and vice versa, serving as a communication layer
- **ROS-I Controller Layer**: This layer consists of vendor-specific industrial robot controllers

So far, we know the basic concepts, so now we are going to discuss how to interface an industrial robot to ROS, using ROS-I packages.

First of all, we will discuss how to develop a **Unified Robot Description Format** (**URDF**) model of an industrial robot and how to create an appropriate MoveIt! configuration for this. Furthermore, we will discuss how to interface the real and simulated Universal Robots and ABB industrial manipulators using ROS-I packages. At the end, we will also discuss the custom IKFast algorithm and plugin development to speed up kinematic computation with MoveIt!.

3D modeling and simulation of an industrial robot and MoveIt!

In this section, we will discuss how to create an URDF file for an industrial robot. We will go through some basic steps on how to create an URDF file for an industrial robot.

As we already know, the URDF is an XML format file that is used to represent a robot model. The URDF file of a robot is the file that describes everything about it, from the visual part and how it looks, to the physical part such as the collision data or the inertia data. URDF files are widely used for representing robots in regular ROS systems. In the case of ROS-I, they are also used for representing industrial robots, but with some peculiarities that we are going to check during this section. Keep in mind that we are not going to learn the fundamentals about URDF files during this section, since we have already discussed them in Chapter 8, *The Robotic Arm in ROS*.

Moreover, in this section, we will discuss how to create a MoveIt! package for our industrial robot to perform motion planning.

Getting ready

Although the URDF model for an experimental robot and an industrial robot are identical, in the case of industrial robots, some recommended standards must be followed. The design of a URDF file should be simple, readable, and modular. It would also be nice to have a common design specification for all industrial robots that are produced by various vendors.

URDF modeling for an industrial robot

There are a few guidelines for URDF design and modeling that are followed by ROS-I:

- **Modular Design**: The URDF design should be modularized using xarco macros, which can be used in a complex and large URDF modeling without any difficulty.
- **Reference Frame**: The `base_link` frame should be the first link, and **tool-zero** (**tool0**) should be the end-effector link.
- **Joint Conventions**: The orientation value of each robot joint is limited to single rotation only. For example, out of the three, orientation-roll, pitch, and yaw values, only one value will be used there.
- **Collision Awareness**: The **inverse kinematics** (**IK**) planners used in the industrial robot are collision aware, therefore the URDF model must have a precise collision 3D mesh for each and every link. Although the mesh file, when used for visual purposes, can have a highly detailed design, the mesh files that are used for collision checking are convex hull detailed mesh designs, which are computationally efficient.

How to do it...

As we have discussed in `Chapter 6`, *ROS Modelling and Simulation*, after building the model using a xacro file for the industrial robot, we can convert it to a URDF file and view the robot model in rviz by using the following command:

```
$ rosrun xacro xacro –inorder –o <output_urdf_file> <input_xacro_file>$
check_urdf <urdf_file>
$ rosrun rviz rviz
```

We will use a xacro file for the Motoman Industrial Robots model for our experiment and learning. We can find the xacro file in the `sia10f_description` package in the `chapter10_tutorials/model/` folder.

Fortunately, MoveIt! provides a very nice and easy-to-use GUI, which will help us interact with the robot in order to perform motion planning. However, before being able to actually use MoveIt!, we need to build a package. This package will generate all the configuration and launch files required for using our defined robot (the one we defined in `sia10f_description`) with MoveIt!. In order to generate this package, we just have to follow all the steps described here:

1. First of all, we'll need to launch the **MoveIt Setup Assistant**. We can do that by typing the following command:

    ```
    $ roslaunch moveit_setup_assistant setup_assistant.launch
    ```

    You should be able to see something similar to the following screenshot:

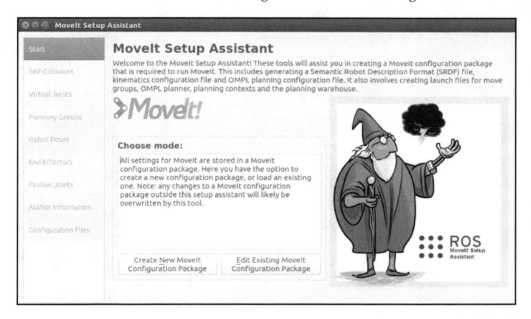

MoveIt Setup Assistant

Great! You can now see the **MoveIt Setup Assistant**. The next thing you'll need to do is load your robot file. So, let's continue:

2. Click on the **Create New MoveIt Configuration Package** button. A new section like this will appear:

Loading a robot description

Now, we will click the **Browse** button, select the xacro file we have for Motoman Industrial Robots in the `sia10f_description` package, and click on the **Load Files** button. We should now see this:

Robot description

Great! We've loaded the xacro file of our robot in to the **MoveIt Setup Assistant**. Now, let's start configuring some things.

Go to the **Self-Collisions** tab and click on the **Regenerate Default Collision Matrix** button.

Here, we are just defining some pairs of links that don't need to be considered when performing collision checking. For instance, because they are adjacent links, they will always be in a collision.

3. Next, we will move to the **Virtual Joints** tab. Here, we will define a virtual joint for the base of the robot and click the **Add Virtual Joint** button before setting the name of this joint to **FixedBase**, and the parent to `world`. Just like this:

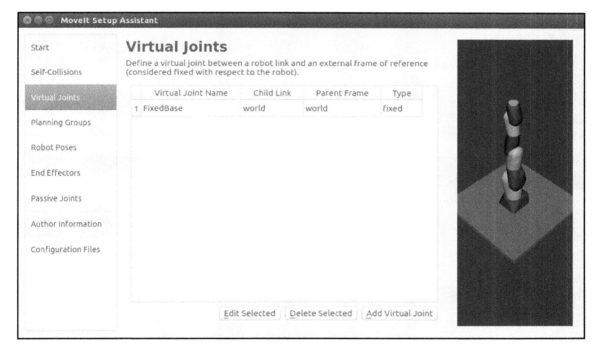

Virtual Joints

4. Finally, we will click the **Save** button. Basically, what we are doing here is creating an "imaginary" joint that will connect the base of our robot with the simulated world.

5. Now, we will open the **Planning Groups** tab and click the **Add Group** button. From here, we will create a new group called `manipulator`, which uses the **KDLKinematics plugin**. Just like this:

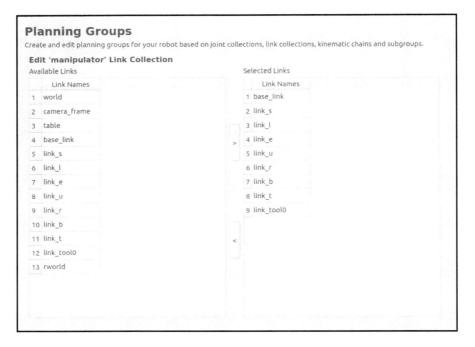

Planning Group

6. Next, we will click on the **Add Kin Chain** button, and we will select the `base_link` as **Base Link**, and the `link_tool0` as **Tip Link**. Just like this:

Kinematic Chain

7. Finally, click the **Save** button and you will end up with something like this:

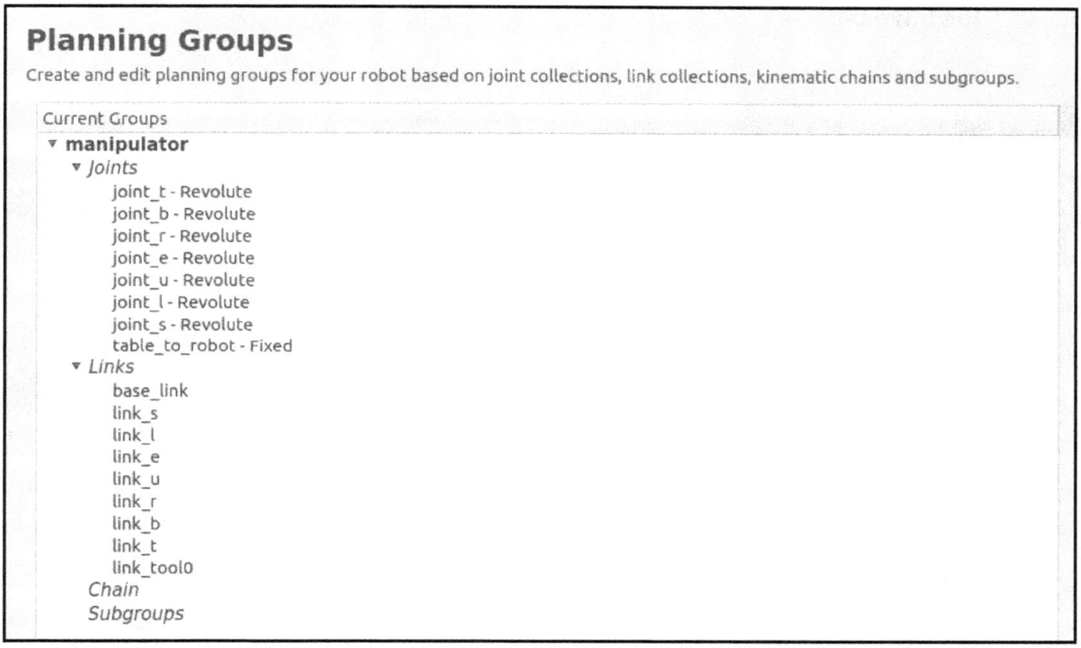

Planning Groups

Create and edit planning groups for your robot based on joint collections, link collections, kinematic chains and subgroups.

Current Groups
- ▾ **manipulator**
  - ▾ *Joints*
    - joint_t - Revolute
    - joint_b - Revolute
    - joint_r - Revolute
    - joint_e - Revolute
    - joint_u - Revolute
    - joint_l - Revolute
    - joint_s - Revolute
    - table_to_robot - Fixed
  - ▾ *Links*
    - base_link
    - link_s
    - link_l
    - link_e
    - link_u
    - link_r
    - link_b
    - link_t
    - link_tool0
  - *Chain*
  - *Subgroups*

Planning Groups completed

So now, we've defined a group of links for performing Motion Planning, and we've defined the plugin we want to use to calculate those plans.

8. Now, we are going to create a couple of predefined poses for our robot. We will go to the **Robot Poses** tab and click on the **Add Pose** button. Our robot will appear with all its joints set to 0. We will name this pose `allZeros`, and click on the **Save** button:

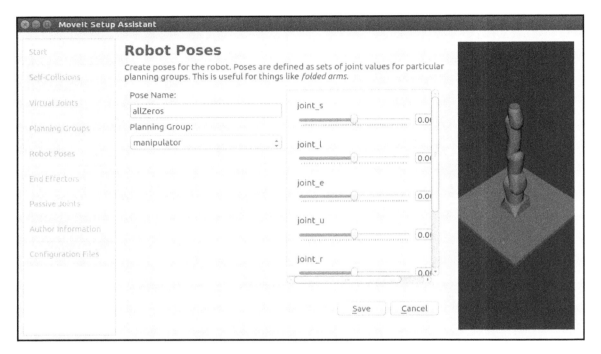

Allzeros configuration

Now, we will repeat this operation, but this time adjusting the position of the joints so that the robot is in a specific position that we will call **home**. We can set the joints to anything we like, but it is recommended to set the robot in a position that is not complicated. For instance, something like this:

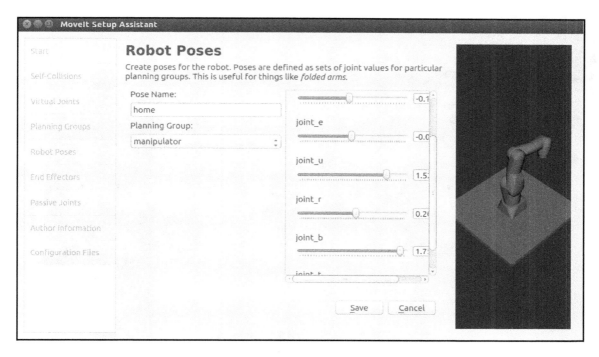

Home configuration

This is very useful, for instance, when we know that there are some poses that our robot will have to go through many times:

Let's enter our name and email in the **Author Information** tab.

9. Finally, we will go to the **Configuration Files** tab and click the **Browse** button; navigate to the workspace directory and create a new directory. Name it xxx_moveit_config and **Choose** the directory we've just created:

Configuration package

10. Click the **Generate Package** button. If everything goes well, we should see something like this:

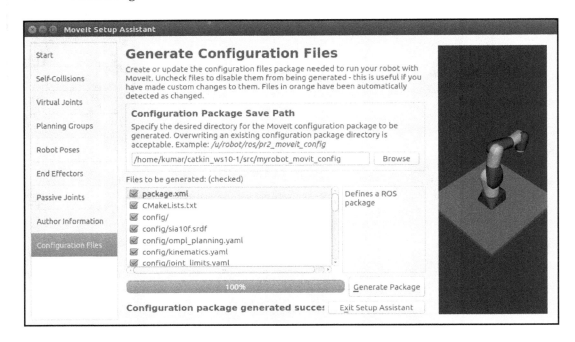

Configuration Packages completed

And that's it! We have just created a MoveIt! package for our industrial robot.

Well… to start, we can just launch the MoveIt! rviz environment and begin doing some experiments regarding Motion Planning, as shown in the following screenshot:

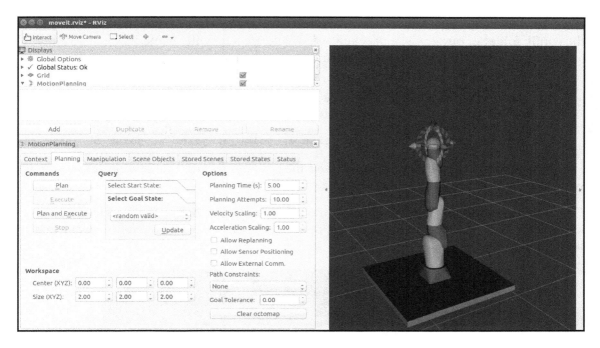

Motion Planning

Controlling the robot in the simulation

Until now, we've only moved the robot in the MoveIt! application. This is very useful because we can do many experiments without worrying about any damage. Anyway, the final goal will always be to move the real robot, right?

The MoveIt! package we've created is able to provide the necessary ROS services and actions in order to plan and execute trajectories, but it isn't able to pass these trajectories to the real robot. All the kinematics we've been performing are executed in an internal simulator that MoveIt! provides. In order to communicate with the real robot, it will be necessary to do a couple of modifications to the MoveIt! package we created at the beginning of this section.

Obviously, we may not have a real robot to do this, so we will apply the same but for moving the simulated robot. We will look at what we need to change in our MoveIt! package in the following subsection.

We'll need to create a file to define how we will control the joints of our real robot. Inside the config folder of our `moveit` package, we will create a new file named `controllers.yaml` and copy the following content inside it:

```
controller_list:
  - name: sia10f/joint_trajectory_controller
    action_ns: "follow_joint_trajectory"
    type: FollowJointTrajectory
    joints: [joint_s, joint_l, joint_e, joint_u, joint_r, joint_b, joint_t]
```

Here, we are defining the Action Server (and the type of message that it will use), which we will be using to control the joints of our robot.

First, we are setting the name of our joint trajectory controller Action Server. How do we know that? Well, if we do a `rostopic` list in any terminal, we'll find the following structure between our topics:

```
/sia10f/joint_trajectory_controller/follow_joint_trajectory/cancel
/sia10f/joint_trajectory_controller/follow_joint_trajectory/feedback
/sia10f/joint_trajectory_controller/follow_joint_trajectory/goal
/sia10f/joint_trajectory_controller/follow_joint_trajectory/result
/sia10f/joint_trajectory_controller/follow_joint_trajectory/status
```

Controller

This way, we know that our robot has a joint trajectory controller Action Server that is called `/sia10f/joint_trajectory_controller/follow_joint_trajectory/`.

We can also find this out by checking the message that this action uses, and that it is of the type `FollowJointTrajectory`.

Finally, we already know the names of the joints that our robot uses. We saw them while we were creating the MoveIt! package, and we can also find them in the `sia10f_macro.xacro` file, in the `sia10f_description` package.

Next, we'll have to create a file to define the names of the joints of our robot. Again, inside the `config` directory, we will create a new file called `joint_names.yaml`, and copy the following content into it:

```
controller_joint_names: [joint_s, joint_l, joint_e, joint_u, joint_r, joint_b, joint_t]
```

Now, if we open the `xxx_moveit_controller_manager.launch.xml` file, which is inside the launch directory, we'll see that it's empty. We will put the following content inside it:

```
<launch>
  <rosparam file="$(find myrobot_moveit_config)/config/controllers.yaml"/>
  <param name="use_controller_manager" value="false"/>
  <param name="trajectory_execution/execution_duration_monitoring"
value="false"/>
  <param name="moveit_controller_manager"
value="moveit_simple_controller_manager/MoveItSimpleControllerManager"/>
</launch>
```

What we are doing here is basically loading the `controllers.yaml` file we have just created and the `MoveItSimpleControllerManager` plugin, which will allow us to send the plans calculated in MoveIt! to our real robot, which in this case is the simulated robot.

Finally, we will have to create a new launch file that sets up all the systems to control our robot. So, inside the launch directory, we will create a new launch file called `xxx_planning_exeution.launch`:

```
<launch>

  <rosparam command="load" file="$(find
myrobot_moveit_config)/config/joint_names.yaml"/>

  <include file="$(find
myrobot_moveit_config)/launch/planning_context.launch" >
    <arg name="load_robot_description" value="true" />
  </include>

  <node name="joint_state_publisher" pkg="joint_state_publisher"
type="joint_state_publisher">
    <param name="/use_gui" value="false"/>
    <rosparam param="/source_list">[/sia10f/joint_states]</rosparam>
  </node>

  <include file="$(find myrobot_moveit_config)/launch/move_group.launch">
    <arg name="publish_monitored_planning_scene" value="true" />
  </include>

  <include file="$(find myrobot_moveit_config)/launch/moveit_rviz.launch">
    <arg name="config" value="true"/>
  </include>

</launch>
```

Here, we are loading the `joint_names.yaml` file and launching some launch files we need in order to set up the MoveIt! environment. We could check what those launch files do, if we want, but let's focus on the `joint_state_publisher` node that is being launched for the moment.

If we create a `rostopic` list again, we will see that there is a topic called `/sia10f/joint_states`. It in this topic that the state of the joints of the simulated robot are published. So, we need to put this topic in the `/source_list` parameter so that MoveIt! knows where the robot is at each moment.

Finally, we just have to launch this file and plan a trajectory, just like we learned to do in the previous section. Once the trajectory is planned, we can press the **Execute** button in order to execute the trajectory in the simulated robot. This is shown in the following screenshot:

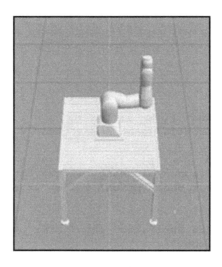

Robot simulation

Great! Now we know how to interact with an industrial robot using the MoveIt! rviz application, which is a very interesting and useful tool.

Working with ROS-I packages – Universal Robots, ABB robot

In this section, we are going to work with two of the most popular ROS-I packages—ABB robots and Universal Robots. We will install the ROS-I packages for both the vendor and manufacturer and work with the MovetIt! interface to simulate industrial robots in Gazebo.

Getting ready

Universal Robots (http://www.universal-robots.com/) is an industrial robot manufacturer based in Denmark. The company mainly produces three arms–UR3, UR5, and UR10, which are shown in the following screenshot:

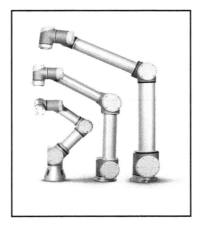

Universal Robots

The specifications of these robots are given in the following table:

Robot Model	UR-3	UR-5	UR-10
Working Space (radius)	500 mm	850 mm	1,300 mm
Payload	3 kg	5 kg	10 kg
Weight	11 kg	18.4 kg	28.9 kg
Footprint	118 mm	149 mm	190 mm

Similarly, ABB Robotics (http://new.abb.com/products/robotics) is a world-leading manufacturer of industrial robots and robot systems that's operating in 53 countries, in over 100 locations around the world. However, we will work with two of the most popular ABB industrial robot models—IRB 2400 and IRB 6640—which are shown in the following photograph:

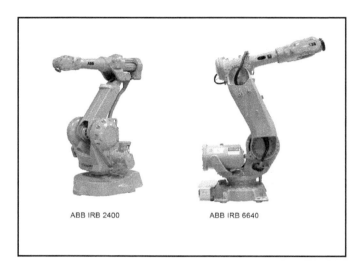

ABB Robotics

The specifications of these robots are given in the following table:

Robot	IRB 2400-10	IRB 6640-130
Working space (radius)	1.55 m	3.2 m
Payload	12 kg	130 kg
Weight	380 kg	1,310-1,405 kg
Footprint	723x600 mm	1,107 x 720 mm

Universal Robots

We can install the Universal Robots packages by using the Debian package manager. Alternatively, we can directly download these packages from the repository, as follows:

```
$ sudo apt-get install ros-kinetic-universal-robot
$ git clone https://github.com/ros-industrial/universal_robot.git
```

After compilation or installation, we can launch the simulation in Gazebo for the UR-10 robot by using the following command:

```
$ roslaunch ur_gazebo ur10.launch
```

We will see something similar to the following screenshot:

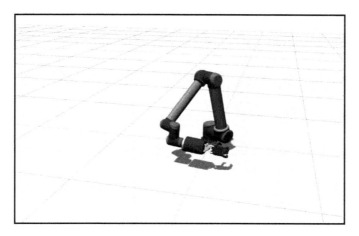

Universal Robot simulation in Gazebo

The Universal Robot packages consist of the following packages:

- `ur_description`: This package consists of the robot description URDF model for UR-3, UR-5, and UR-1
- `ur_driver`: This package contains driver nodes, which communicate with the hardware controllers of the UR-3, UR-5, and UR-10 robots
- `ur_bringup`: This package consists of launch files so that we can start working with the real robot
- `ur_gazebo`: This package consists of Gazebo simulations of UR-3, UR-5, and UR-10
- `ur_msgs`: This package contains ROS message files which are used for communication among various nodes of Universal Robot packages
- `urXX_moveit_config`: This packages contains MoveIt config files of Universal Robot manipulators—`ur3_moveit_config`, `ur5_moveit_config`, and `ur10_moveit_config`
- `ur_kinematics`: This package contains Kinematic solver plugins for the Universal Robot model, which can be also used with MoveIt!

Furthermore, we can refer to the robot controller configuration file in the ur_gazebo/controller folder, which is used for interfacing with the MoveIt! package. The MoveIt! configuration for the Universal Robot can be found in the config directory of each moveit_config package for UR-3, UR-5, and UR-10.

In the same directory, we can find the Kinematic configuration file kinematics.yaml which specifies the IK solvers used for the specific robotic arm model. The content of the Kinematic configuration file for the UR-10 robot model is shown as follows:

```
#manipulator:
#   kinematics_solver: ur_kinematics/UR10KinematicsPlugin
#   kinematics_solver_search_resolution: 0.005
#   kinematics_solver_timeout: 0.005
#   kinematics_solver_attempts: 3
manipulator:
  kinematics_solver: kdl_kinematics_plugin/KDLKinematicsPlugin
  kinematics_solver_search_resolution: 0.005
  kinematics_solver_timeout: 0.005
  kinematics_solver_attempts: 3
```

Likewise, we could also refer to the ur10_moveit_controller_manager.launch file for the UR-10 model and another inside the launch folder which loads the trajectory controller configuration and starts the trajectory controller manager, as shown in the following listing:

```
<launch>
  <rosparam file="$(find ur10_moveit_config)/config/controllers.yaml"/>
  <param name="use_controller_manager" value="false"/>
  <param name="trajectory_execution/execution_duration_monitoring"
value="false"/>
  <param name="moveit_controller_manager"
value="moveit_simple_controller_manager/MoveItSimpleControllerManager"/>
</launch>
```

Moving forward, let's learn how to perform motion planning using MoveIt! and simulate it using Gazebo. As we have discussed in Chapter 8, *The Robot Arm in ROS*, we have to follow the following steps:

1. Start the simulation of UR-10 with joint trajectory controllers:

```
$ roslaunch ur_gazebo ur10.launch
```

2. Start the MoveIt! nodes for motion planning with `sim:=true` to run MoveIt! as a simulation:

```
$ roslaunch ur10_moveit_config
ur10_moveit_planning_execution.launch sim:=true
```

3. Launch rviz with the MoveIt! visualization plugin:

```
$ roslaunch ur10_moveit_config moveit_rviz.launch config:=true
```

We can move the end-effector position of the robot and plan the path by using the **Plan** button. When we press the **Execute** button or the **Plan and Execute** button, the trajectory should be sent to the simulated robot, performing the motion in the Gazebo environment, as shown in the following screenshot:

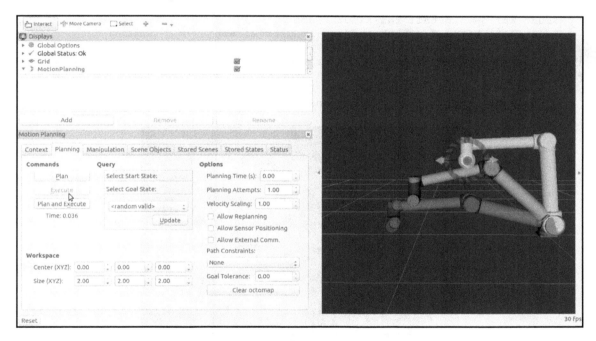

Motion planning in the UR-10 model in rviz

Once our control algorithms are validated in the Gazebo simulation, we can start experimenting with the manipulation tasks with a real Universal Robots arm. The main difference between working with a simulation and a real robot is that we have to start the driver for the hardware controller, which will communicate with the real robot using the provided hardware-software interface.

Although the default driver of Universal Robot arms has been released with the `ur_driver` package of ROS-I, for the newer versions of these systems (v3.x and up), it is recommended that you use the unofficial `ur_modern_driver` package:

```
$ git clone ur10_moveit_config ur10_moveit_planning_execution.launch
sim:=true
```

We can refer to the manual for specific versions of systems that we can work with.

ABB Robots

We can install the ABB robots packages by using the Debian package manager, or we can directly download these packages from the repository as follows:

```
$sudo apt-get install ros-kinetic-abb
$git clone https://github.com/ros-industrial/abb
```

We can launch the ABB IRB 6640 in rviz for motion planning which is shown in screenshot below, using the following command:

```
$ roslaunch abb_irb6640_moveit_config demo.launch
```

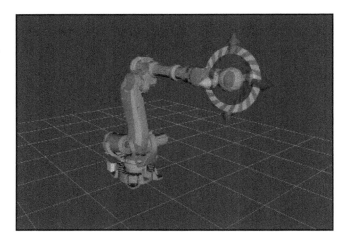

Motion planning of ABB IRB 6640

Similarly, we can also launch the ABB IRB 2400 in rviz for motion planning, which is shown in the screenshot below, using the following command:

```
$ roslaunch abb_irb2400_moveit_config demo.launch
```

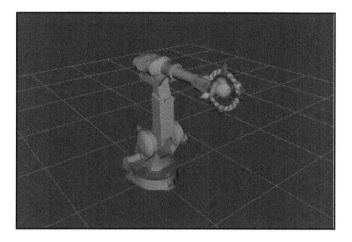

Motion planning of ABB IRB 2400

We can refer to the ABB ROS packages and manuals for more details regarding specification and configuration so that we can work with a particular version of the system.

ROS-I Robot support packages

The ROS-I Robot Support Packages are modern conventions which are pursued for industrial robots. The major goal of these support packages is to standardize the concepts of maintaining ROS packages for a wide variety of industrial robots that are designed by several vendors and then manufactured.

All official supported packages in ROS-I share a common file and directory layout across all models and all manufacturers.

Getting ready

We have already cloned the ABB robot packages in the previous section, and inside the abb folder, we could see several support packages which support varieties of ABB robots, such as:

- abb_irb2400_support
- abb_irb4400_support
- abb_irb5400_support
- abb_irb6600_support
- abb_irb6640_support

Here, we are taking the ABB IRB 2400 model support package called abb_irb2400_support as a case study. The following screenshot shows the top-level directory layout for a robot support package:

config
launch
meshes
tests
urdf
CHANGELOG.rst
CMakeLists.txt
package.xml
readme.md

ABB IRB 2400 support packages

In short, the **config** directory contains files that keep information such as joint names, rviz configurations, or any other model-specific configuration.

Similarly, the **meshes** directory stores all the 3D meshes referenced in the URDFs that are used for visualization and collision detection.

The **launch** directory contains a set of launch files that are used to support all packages. Moreover, each support package also contains a set of standardized (roslaunch) tests, which can be found in the test directory. These tests typically check the launch files during the launch for errors, along with they may perform other tests also.

In addition, all URDFs and xacros are stored in the URDF directory:

- `config`: Inside the `config` folder, there is a configuration file named `joint_names_irb2400.yaml`, which contains the joint names of the robot that are used by the ROS controller.
- `launch`: The `launch` folder contains the `launch` file definitions of the robot. These files follow a common convention used by all industrial robots:

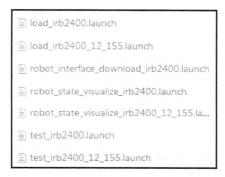

Launch

- `load_irb2400.launch`: This file commonly loads `robot_description` on the parameter server and all xacro files in a single launch file. Therefore, instead of writing separate code for adding `robot_description` in other launch files, we could easily include this `launch` file.
- `test_irb2400.launch`: This `launch` file can be used in visualization of the loaded URDF which includes the preceding launch files and starts the `joint_state_publisher` and `robot_state_publisher` nodes, which help us interact with the user on `rviz`.
- `robot_state_visualize_irb2400.launch`: This launch file can be used in the visualization of the current state of the real robot which runs nodes from the ROS-I driver package with the appropriate parameters. However, this launch file needs a real robot or simulation interface. The current state of the robot is visualized by running rviz and the `robot_state_publisher` node.
- `robot_interface_download_irb2400.launch`: This launch file could be used for bi-directional communication with the industrial robot controller to ROS and vice versa. In addition, this launch file also requires access to the simulation or real robot controller.

Establishing the communication between a real robot controller and ROS requires the IP addresses of the industrial controllers, where the controller should be running the ROS-I server programs.

- urdf: This folder contains a set of standardized xacro files of the robot model:

URDF

- irb2400_macro.xacro: This is the xacro definition of a specific robot.
- irb2400.xacro: This is the top-level xacro file which creates an instance of the macro and doesn't include any other files other than the macro of the robot. This xacro file will be loading inside the load_irb2400.launch file which was discussed in the preceding section and chapter.
- irb2400.urdf: This is the URDF generated from the preceding xacro file using the xacro tool. This file is used when the tools or packages can't load xacro directly, however, this is the top-level URDF for this robot.
- meshes: This contains meshes for visualization and collision checking.
- tests: This folder contains the test launch file to test all the preceding launch files.

We can experiment with the robot model using the test_irb2400.launch file, which will launch the test interface of the ABB IRB 2400 robot:

```
$ roslaunch abb_irb2400_support test_irb2400.launch
```

The preceding command will show the robot model in `rviz` with a joint state publisher node, as shown in the following screenshot:

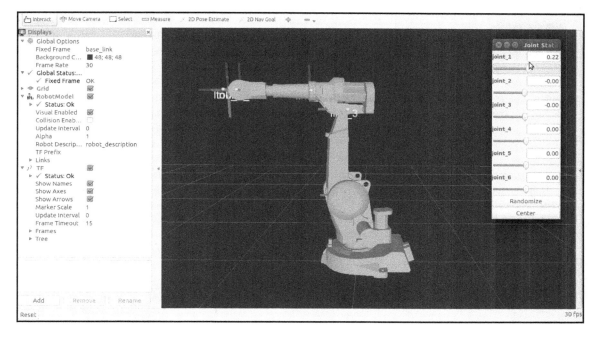

ABB IRB 2400 with Joint State Publisher in rviz

ROS-I Robot client package

The ROS-I Robot Client Package provides a standardized interface for controlling industrial robots which are based on the ROS-I Specification. The industrial robot client communicates to the robot controller using the `simple_message` protocol, where the industrial robot controller must have a server running. The industrial robot client nodes communicate with a compatible server running on a standalone industrial robot controller.

Primarily, this package provides the `industrial_robot_client` library for robot-specific implementations so that we can reuse code from this library using standard C++ derived-class mechanisms. Moreover, this package also provides generic nodes exposing the base `industrial_robot_client` functionality. The following web page describes this usage in more detail: `http://wiki.ros.org/industrial_robot_client/generic_implementation`.

Getting ready

There are differences among manufacturer interfaces and robot designs. These will require changes to the basic reference implementation provided in the `industrial_robot_client` library which includes joint-coupling, velocity scaling, and communications protocols.

It is recommended that the developer uses a derived-class approach and, wherever possible, re-implements the required minimal functionality only. This approach will avoid code duplication and maintain consistent operations.

It may be helpful to refer to the library's design at `http://wiki.ros.org/industrial_robot_client/design`. However, simple joint reordering and renaming can be handled through existing capabilities, as described at `http://wiki.ros.org/industrial_robot_client/joint_naming`.

In the following example, we will discuss how to replace the reference velocity calculation (0—100% of max-velocity) with an absolute velocity algorithm.

First of all, we will create a new derived class, based on `JointTrajectoryDownloader`:

```
class myRobot_JointTrajectoryDownloader : JointTrajectoryDownloader
{
...
}
```

Next, we will override the specific functionality. In this case, this is the velocity calculation:

```
class myRobot_JointTrajectoryDownloader : JointTrajectoryDownloader
{

public:
  bool calc_velocity(const trajectory_msgs::JointTrajectoryPoint& pt,
double* rbt_velocity)
  {
    if (pt.velocities.empty())
      *rbt_velocity = SAFE_SPEED;
    else
      *rbt_velocity = std::min(pt.velocities.begin(), pt.velocities.end());

    return true;
  }
}
```

Finally, we will implement a new node using an instance of the appropriate derived-class:

```
int main(int argc, char** argv)
{
  // initialize node
  ros::init(argc, argv, "motion_interface");

  myRobot_JointTrajectoryDownloader motionInterface;
  motionInterface.init();
  motionInterface.run();

  return 0;
}
```

Furthermore, we can also refer to the examples of creating robot-specific interfaces for the ABB driver client at https://github.com/ros-industrial/abb/blob/kinetic-devel/abb_driver/src/abb_joint_downloader_node.cpp.

ROS-I robot driver specification

In this section, we will discuss the specification provided by the ROS-I Consortium for ROS node functionality to improve cross-platform compatibility. All ROS nodes that provide an interface to an industrial robot controller should follow the specification.

We can send the comments and questions on this specification to the ROS-I mailing list (swri-ros-pkg-dev@googlegroups.com).

In the following section, we will look into an overview of the guidelines that are provided by the ROS-I Consortium.

Getting ready

One of the primary goals of ROS-I is facilitating interoperability among robots from different vendors by integrating their control with the common ROS framework. These different ROS control nodes all utilize a common set of interfaces for control and feedback. Also, ROS-I has provided the guidelines for specific ROS interfaces to ensure maximum compatibility.

First of all, we will discuss how the robot should be expected to respond to high-level operational activities:

- **Initialization**: The ROS node should automatically initialize all connections to the robot controller, and the robot startup should make the controller program run automatically.
- **Communications**: Both the ROS node and the robot controller program must handle the communications-loss scenarios so that they can implement a heartbeat message between them.

Next, we are going to discuss specific ROS interfaces—topics, services, and parameters. These interfaces must be provided by the robot controller program, which might be a single node or multiple nodes, as required by the robot communications architecture. The robot controller program must incorporate the following parameters:

- `robot_ip_address` (string): IP address of the robot
- `robot_description` (urdp map xml): The URDF description of the robot

Further, a robot controller program should publish the following topics:

- `feedback_states` (`control_msgs/FollowJointTrajectory`): Provides feedback on the current and desired joint state, which is used by `joint_trajectory_action`
- `joint_states` (`sensor_msgs/JointState`): Provides feedback on the current joint state, which is used by the `robot_state_publisher` node to broadcast kinematic transforms
- `robot_status` (`industrial_msgs/RobotStatus`): Provides the current status of the robot, which is used by application code to monitor and react to different fault conditions

In addition, the robot controller program which controls the robot's movement should subscribe to the following topics:

- `joint_path_command` (`trajectory_msgs/JointTrajectory`): Executes a pre-calculated joint trajectory on the robot and is used by ROS's trajectory generator's `joint_trajectory_action` to issue motion commands
- `joint_command` (`trajectory_msgs/JointTrajectoryPoint`): Executes a dynamic motion by streaming joint commands while in motion, which is used by client code to control the robot's position in real time

Finally, there should be service interfaces provided by the robot controller such as:

- `stop_motion(industrial_msgs/StopMotion)`: Stop current robot motion
- `joint_path_command(industrial_msgs/CmdJointTrajectory)`: Execute a new motion trajectory on the robot

To implement real-time motion planning and collision avoidance, the node should provide robot-specific inverse kinematics (IK) solutions. However, a generic numerical solver is provided in ROS that works with most cases, but it operates too slowly for collision-avoidance path planning. It is possible to have a custom version of IK-solver (`http://wiki.ros.org/Industrial/Tutorials/Create_a_Fast_IK_Solution`). In the next section, we will discuss how to generate an IK solver plugin using IKFast, a powerful inverse kinematics solver.

The ROS path planners and collision checkers implement a high-order smoothing algorithm for trajectories following waypoints. However, the resulting trajectory, as executed by the robot controller, might not exactly match the *ideal* planned trajectory. Therefore, the ROS path planners add a specified amount of *padding* to the robot models to account for differences between the planned and actual paths. The amount of *padding* has to be calculated experimentally, which results in collision-free motion mostly; however it increases the level of padding and reduces the chance of collision.

Although appropriate integration between path planners and robot controllers would reduce both the path-execution error and the collision-padding requirement, it requires more effort from specific robot controller developers.

As discussed in the previous section, a reference implementation of an ROS-I node implementing these capabilities can be seen at `http://wiki.ros.org/industrial_robot_client`. This uses a simple message-based socket protocol to communicate with the industrial robot controller. We can also refer to the ABB and Motoman implementations for examples, which have been described in the previous section. The example programs at `abb_driver or ur_driver` describe how to integrate a robot-side server application with an ROS-I client node derived from these reference implementations.

Developing a custom MoveIt! IKFast plugin

In the preceding section, we have learned how to configure the MoveIt! packages with a particular robotic arm to perform motion planning. The MoveIt! package use the default KDL kinematics plugin, which finds IK solutions using a numerical method.

This default numerical IK solver, KDL, is mainly used so that robots have a degree of freedom (DOF > 6), whereas most of the industrial robotic arms have DOF <= 6. Therefore, we can use analytic solvers, which are much faster than numerical solvers.

Getting ready

In this section, we will discuss how to generate an IK solver plugin using IKFast, a powerful inverse kinematics solver provided by the OpenRAVE (http://openrave.org/) motion planning software.

A good example of a robot using IKFast in ROS is the Motoman sia10d, universal robots UR5 and UR10, and ABB robotics IRB 2400.

OpenRAVE installation

The best solution to install OpenRAVE today is to build it from source, which fortunately is not so difficult. The instructions in the following section are for Ubuntu 16.04 and 18.04.

First, we have to make sure that the following programs are installed on our system:

```
$ sudo apt-get install cmake g++ git ipython minizip python-dev python-h5py
python-numpy python-scipy python-sympy qt4-dev-tools
```

Next, we will have to to install the following libraries from the official Ubuntu repository:

```
$ sudo apt-get install libassimp-dev libavcodec-dev libavformat-dev
libavformat-dev libboost-all-dev libboost-date-time-dev libbullet-dev
libfaac-dev libglew-dev libgsm1-dev liblapack-dev liblog4cxx-dev libmpfr-
dev libode-dev libogg-dev libpcrecpp0v5 libpcre3-dev libqhull-dev libqt4-
dev libsoqt-dev-common libsoqt4-dev libswscale-dev libswscale-dev
libvorbis-dev libx264-dev libxml2-dev libxvidcore-dev
```

The next dependency is collada-dom, which we can clone from GitHub and build from source as well:

```
$ git clone https://github.com/rdiankov/collada-dom.git
$ cd collada-dom && mkdir build && cd build
$ cmake ..
$ make -j4
$ sudo make install
```

Another dependency that we'll be using is OpenSceneGraph:

```
$ sudo apt-get install libcairo2-dev libjasper-dev libpoppler-glib dev
libsdl2-dev libtiff5-dev libxrandr-dev
$ git clone https://github.com/openscenegraph/OpenSceneGraph.git
$ cd OpenSceneGraph && mkdir build && cd build
$ cmake ..
$ make -j4
$ sudo make install
```

In new versions, the OpenRAVE defaults also require that we install the Flexible Collision Library:

```
$ sudo apt-get install libccd-dev
$ git clone https://github.com/flexible-collision-library/fcl.git
$ cd fcl
$ mkdir build && cd build
$ cmake ..
$ make -j4
$ sudo make install
```

Once all of the software is installed, we have to clone the `latest_stable` branch of OpenRAVE from GitHub:

```
$ git clone --branch latest_stable https://github.com/rdiankov/openrave.git
$ cd openrave && mkdir build && cd build
$ cmake .. -DOSG_DIR=/usr/local/lib64/
$ make -j4
$ sudo make install
```

Finally, we have to add OpenRAVE to our Python path by adding these two lines in `.bashrc` or `.zshrc` to save this configuration between sessions:

```
$ export LD_LIBRARY_PATH=$LD_LIBRARY_PATH:$(openrave-config --python-
dir)/openravepy/_openravepy_
$ export PYTHONPATH=$PYTHONPATH:$(openrave-config --python-dir)
```

We can check that our installation works by running one of the default examples:

```
$ openrave.py --example graspplanning
```

This command should fire up the grasp planning example, as shown in the following screenshot:

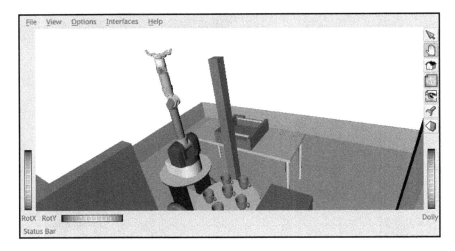

OpenRAVE grasp planning

The IKFast routine requires a model of the robot in either OpenRave's custom XML format or a Collada (https://www.khronos.org/collada/) DAE format. For most robots, the easiest option may be to convert an existing URDF model into the Collada format.

If our model is in `xacro` format, we can convert it to pure URDF:

```
$ rosrun xacro xacro.py my_robot.urdf.xacro > my_robot.urdf
```

Next, we will convert URDF into `collada` format:

```
$ rosrun collada_urdf urdf_to_collada my_robot.urdf my_robot.dae
```

Since we have installed the full version of OpenRAVE, we can view our model:

```
$ openrave my_robot.dae
```

For example, we can open the `irb6640.dae` file for the ABB IRB 2400 robotic arm using OpenRave with the following command:

```
$ openrave irb6640.dae
```

We will see the model in OpenRave, as shown in the following screenshot:

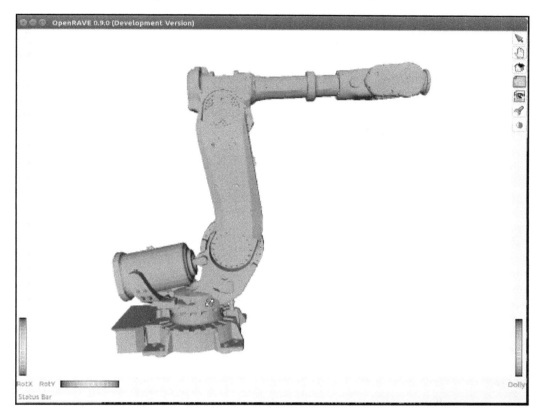

ABB IRB 6640 model with OpenRAVE

How to do it...

We can choose which sort of IK would be appropriate for our model by referring to the following
page: `http://openrave.org/docs/latest_stable/openravepy/ikfast/#ik-types`.
However, the most common IK type is `transform6d`.

We have to provide link index numbers for the `base_link` and `end_link` between which the IK will be calculated. We can count the number of links by checking them in the `.dae` file. Alternatively, if we have OpenRAVE installed, we can view a list of links in our model:

```
$ openrave-robot.py my_robot.dae --info links
```

As usual, the 6-DOF manipulator should have six arm links and a dummy `base_link`, as required by ROS specifications. If there are no extra links present in the model, this gives `baselink=0` and `eelink=6`. Generally, an additional `tool_link` will be provided to position the `grasp-tool` frame, setting it as `eelink=7`.

The following manipulator also has another dummy `mounting_link`, providing `baselink=1` and `eelink=8`:

name	index	parents
base_link	0	
mounting_link	1	base_link
link1_rotate	2	mounting_link
link2	3	link1_rotate
link3	4	link2
link4	5	link3
link5	6	link4
link6_wrist	7	link5
tool_link	8	link6_wrist

Manipulator link number

To generate the IK solution between the manipulator's base and tool frames for a 6 DOF arm, we can use the following command format:

```
$ python `openrave-config --python-dir`/openravepy/_openravepy_/ikfast.py --robot=<myrobot_name>.dae --iktype=transform6d --baselink=1 --eelink=8 --savefile=<ikfast_output_path>
```

`<ikfast_output_path>` is recommended to be a path that points to a file named `ikfast61_<planning_group_name>.cpp`.

Similarly, for a 7 DOF arm, we would have to specify a free link:

```
$ python `openrave-config --python-dir`/openravepy/_openravepy_/ikfast.py --robot=<myrobot_name>.dae --iktype=transform6d --baselink=1 --eelink=8 --freeindex=4 --savefile=<ikfast_output_path>
```

However, the speed and success of this process will depend on the complexity of our robot.

 A typical 6 DOF manipulator with three intersecting axes at the base or wrist will take only a few minutes to generate the IK.

Moreover, we can consult the OpenRAVE mailing list and ROS Answers for information about 5 and 7 DOF manipulators.

Next, we will create the package that will contain the IK plugin. We will name the package `<myrobot_name>_ikfast_<planning_group_name>_plugin`, so that we can refer to our IKFast package simply as `<moveit_ik_plugin_pkg>` later:

```
$ cd ~/catkin_ws/src
$ catkin_create_pkg <moveit_ik_plugin_pkg>
```

Now, we will build our workspace so that the new package is detected:

```
$ cd ~/catkin_ws
$ catkin_make
```

Next, we will create the plugin source code:

```
$ rosrun moveit_ikfast create_ikfast_moveit_plugin.py <myrobot_name>
<planning_group_name> <moveit_ik_plugin_pkg> <ikfast_output_path>
```

We use the following parameters:

- `myrobot_name`: Name of the robot as in our URDF
- `planning_group_name`: Name of the planning group, as referenced in our `kinematics.yaml` file
- `moveit_ik_plugin_pkg`: Name of the new package we have just created
- `ikfast_output_path`: File path to the location where we have generated the IKFast `output.cpp` file

This command will generate a new source file called `<myrobot_name>_<planning_group_name>_ikfast_moveit_plugin.cpp` in the `src/` directory, and modify various configuration files.

Finally, we will build our workspace again to create the IK plugin:

```
$ cd ~/catkin_ws
$ catkin_make
```

This will build the new plugin library
`lib/lib<myrobot_name>_<planning_group_name>_moveit_ikfast_moveit_plugin`
which can be used with MoveIt!.

The IKFast plugin should function identically to the default KDL IK Solver, but with greatly increased performance. The MoveIt! configuration file is automatically edited by the `moveit_ikfast` script, but we can switch between the KDL and IKFast solvers using the `kinematics_solver` parameter in the robot's `kinematics.yaml` file. For example:

```
$ rosed <myrobot_name>_moveit_config/config/kinematics.yaml
```

Edit these parts:

```
<planning_group_name>:
  kinematics_solver: <moveit_ik_plugin_pkg>/IKFastKinematicsPlugin
-OR-
  kinematics_solver: kdl_kinematics_plugin/KDLKinematicsPlugin
```

Moreover, we can use the MoveIt! rviz Motion Planning plugin and use the interactive markers to see if the correct IK solutions are generated.

ROS-I-MTConnect integration

Like most businesses, manufacturing shop floors are constantly trying to increase productivity, profitability, and efficiency. However, the modern manufacturing shop floor contains many different types of machining equipment, each supporting different proprietary interfaces and communication protocols. This diversity has made it extremely challenging for manufacturers to monitor and maintain their machines.

Fortunately, the emergence of MTConnect has made it possible for a machine monitoring system to consistently and accurately collect data from any MTConnect-compatible machine. This is shown in the following diagram:

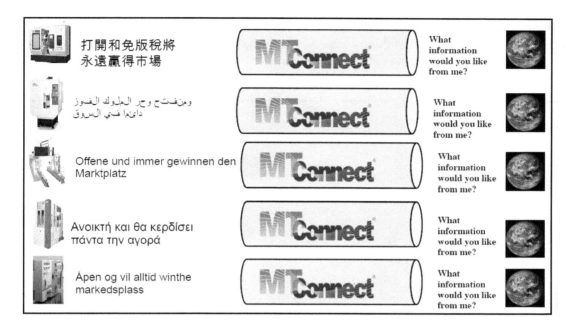

The manufacturing world after MTConnect

Getting ready

MTConnect is not an application. The MTConnect Institute (http://www.mtconnect.org/) does not sell MTConnect.

MTConnect enables end-user productivity by:

- Making it easy to connect manufacturing equipment to a network and to get information from them
- Making it easy to share manufacturing information in an open and standard way with their suppliers anywhere they choose
- Making it very easy to monitor what other manufacturing equipment is doing
- Allowing us to easily analyze all the information in our plant
- Using MTConnect—connect once and improve productivity everywhere

MTConnect is a standard based on open and royalty free protocols (a language with a dictionary) that enables manufacturing equipment to speak in the language of the internet. The language of the internet is HTTP and XML, which is the language of all browsers. The following diagram shows peer-to-Peer MTConnect messaging:

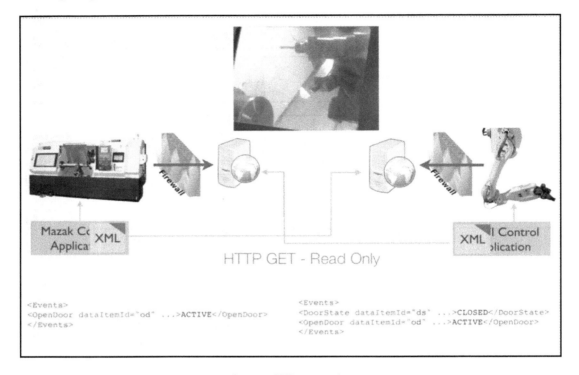

Peer-to-peer MTConnect messaging

MTConnect makes manufacturing equipment access as easy as any website on internet with our browser, for example, `http://MyMachineTool.MyPlant.com`.

MTConnect is like Bluetooth, which allows different devices to easily speak to each other in a common language, where the translation unit resembles the MTConnect protocols, as shown in the following diagram:

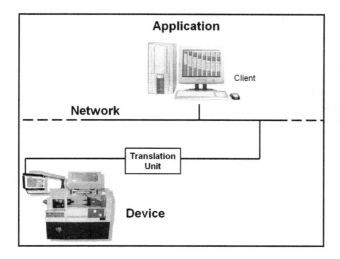

Application

Client

Network

Translation
Unit

Device

Manufacturing equipment on the internet

How to do it...

The ROS-I-MTConnect Integration program has the goal of developing a bridge between MTConnect and ROS-I. The MTConnect software stack contains packages and libraries for integrating ROS and equipment or machine tools that support the MTConnect communications standard in a manufacturing environment. Similar to ROS messaging, the MTConnect communications standard describes both the semantic data definition and method of communication.

This stack is part of the ROS Industrial and MTConnect programs. It contains packages that create a bridge between ROS and the MTConnect protocols.

The installation of this stack requires both binary installations and installations from the source.

Ruby and its associated state machine library are required to run some scripts, so we have to install them:

```
$ sudo apt-get install ruby
$ sudo gem install statemachine
```

MTConnect ROS bridge and MTConnect Agent libraries will all be downloaded from the source directory as follows:

```
$ git clone git://github.com/mtconnect/ros_bridge.git
$ git clone git://github.com/mtconnect/cppagent.git
$ mkdir agent_build
$ cd agent_build
$ cmake ../cppagent
$ make
```

The following environment variables must be defined in the ~/.bashrc file:

```
$ export ROS_PACKAGE_PATH=<source path>:$ROS_PACKAGE_PATH
$ export MTCONNECT_AGENT_DIR=<source path>/agent_build/agent/
```

The MTConnect example stack contains sample integrations between a simulated CNC and robot. The following command will start the ROS bridge, MTConnect agents, and CNC simulator:

```
$ roslaunch mtconnect_ros_bridge mtconnect_ros_bridge_components.launch
```

We can see that multiple terminal windows will be opened for the various agents and CNC simulator.

Future of ROS-I – hardware support, capabilities, and applications

In this section, we will talk about what can we do with ROS-I today. We will focus our discussion section on ROS deployment or ROS being deployed in an industrial setting.

We will talk about the lowest components of the POS-I software stack, the hardware devices, ROS capabilities in an industrial setup, and amazing applications.

Currently, ROS-I has a device driver for five different robot manufactures: ABB, Fanuc, Universal Robots, Motoman, and Adept, which are shown in the following photograph:

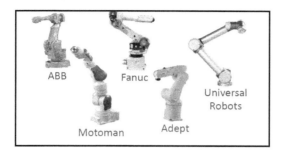

ROS Industrial robot manufactures

There are a couple of models from these robot manufactures that are supported in ROS-I packages. The new robot models can be easily added by these manufacturers, which do not require any software changes, expect configuration. Sometimes, robot mesh and kinematics need to be added and configured.

ROS-I also provides the framework to add new robot controllers, which requires the development of the robot-side driver/server and ROS client modifications (if needed); however, several templates are available, as shown in the following diagram:

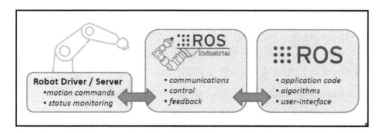

New robot controller development

ROS-I provides a simulation environment for any existing robot model for code development evaluation, such as MoveIt! packages. Besides the robot models, there is strong support for perception sensors, which includes the 2D sensors-camera, SICK laser rangefinder, 3D sensor-stereo vision, Kinect, swissranger, and many more, as shown in the following photograph:

Perception sensors

From an I/O prospective, the current setup mostly works with the EthernetCAT I/O rack, which is a little bit limited.

As discussed in the previous section, ROS-I has a ROS/MTConnect bridge for communication and data transfer, which is mostly used for synchronizing a robot with CNC operations, as shown in the following diagram:

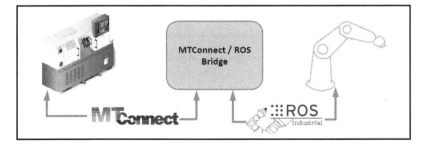

ROS/MTConnect Bridge

However, the current ROS-I setup does not support PLC Interfaces, Industrial Networks-Ethernet/IP, DeviceNet, Modbus, Profibus, OPC, and Industrial HMIs. These are capabilities which are on the road-map vision and would be nice to add in ROS-I, depending upon their priority.

Using the perception sensors, ROS-I has a ton of capabilities to perceive the environment for making robots autonomous. Some of them require segmentation and identification so that the robots know what and how to pick up and grasp objects. The mobile platform robot is provided with mapping and localization capabilities in a 2D and 3D environment, as shown in the following picture:

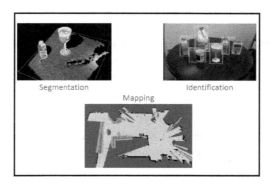

Perception capabilities

We can see lots of planning capabilities for ROS-I, which includes *collision-aware path planning* to automatically find a path to a target position and avoid obstacles, joint limits, and so on with fast collision checking in real time. As you can see from the following photograph, the robot is trying to get some objects from the closet:

Collision-aware path planning

We have seen a strong visualization environment which is the key strength of offline development, which provides review path-plans, debugs control logic, and develops applications even with no hardware. This enables a smooth transition to the physical system. This has the limitation of not having a full physics-based simulation available, but Gazebo physic engines are most encouraging.

It also has the capability of recording and replaying data for both sensor and control data, which adds strength for development and troubleshooting.

We can use the same offline visualization tools for online visualization processes as well. This also provides visualization updates based on live sensor feedback for remote process monitoring and "augmented reality" displays.

Moreover, ROS-I also provides some level of hardware independence—running the same program on different robot platforms which helps in reducing software "porting" problems and allows us to select robots based on capabilities, rather than familiarity.

We will also discuss some of the missing capabilities of ROS-I. Some of them are precision control of path pos/vel/timing, and simplified programming interfaces such as the teach panel interface for a factory floor person. Other limitations include dynamic path modification and motion-related triggers to adjust the path's offset. These are potential future enhancements.

We will also look at the ROS-I from an applications perspective. A few applications include unknown part stream and unstructured picking. These applications use ROS-I technologies such as 3D scanning, segmentation, filtering, and dynamic path planning; however, this is not possible with a traditional industrial robot. The following photograph shows the part picking applications:

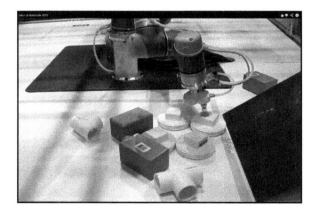

Part picking application

One of the hard application areas supported by ROS-I is a mobile robot with a robotic arm for inventory management and large workspaces, which uses core ROS technologies such as base/arm support, mapping, localization, path planning, and navigation. One such research platform is the PR2 robot in a university and research institute. Some of the other application areas include a dynamic environment, low downtime, and heterogeneous workspaces:

Less industrial applications

We could not resist discussing some of the amazing less industrial applications such as cooking breakfast, folding laundry, playing pool, baking cookies, fetching beer, scooping poop, and many more, which are shown in the preceding photograph.

Other Books You May Enjoy

If you enjoyed this book, you may be interested in these other books by Packt:

ROS Robotics Projects
Lentin Joseph

ISBN: 978-1-78355-471-3

- Create your own self-driving car using ROS
- Build an intelligent robotic application using deep learning and ROS
- Master 3D object recognition
- Control a robot using virtual reality and ROS
- Build your own AI chatter-bot using ROS
- Get to know all about the autonomous navigation of robots using ROS
- Understand face detection and tracking using ROS
- Get to grips with teleoperating robots using hand gestures
- Build ROS-based applications using Matlab and Android
- Build interactive applications using TurtleBot

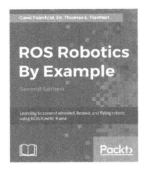

ROS Robotics By Example - Second Edition

Corey P. Schultz, Bob Perciaccante

ISBN: 978-1-78847-959-2

- Control a robot without requiring a PhD in robotics
- Simulate and control a robot arm
- Control a flying robot
- Send your robot on an independent mission
- Learning how to control your own robots with external devices
- Program applications running on your robot
- Extend ROS itself
- Extend ROS with the MATLAB Robotics System Toolbox

Leave a review - let other readers know what you think

Please share your thoughts on this book with others by leaving a review on the site that you bought it from. If you purchased the book from Amazon, please leave us an honest review on this book's Amazon page. This is vital so that other potential readers can see and use your unbiased opinion to make purchasing decisions, we can understand what our customers think about our products, and our authors can see your feedback on the title that they have worked with Packt to create. It will only take a few minutes of your time, but is valuable to other potential customers, our authors, and Packt. Thank you!

Index

www.ingramcontent.com/pod-product-compliance
Lightning Source LLC
Chambersburg PA
CBHW060643060326
40690CB00020B/4497